Richard Béliveau, PhD ▪ **Denis Gingras**, PhD

Death

The Scientific Facts to Help Us Understand It Better

FIREFLY BOOKS

Published by Firefly Books Ltd. 2012 under arrangement with Groupe Librex inc. doing business under the name of Éditions du Trécarré, Montréal, QC, Canada

First printing

Publisher Cataloging-in-Publication Data (U.S.)

Béliveau, Richard.
Death : the scientific facts to help us understand it better / Richard Béliveau and Denis Gingras.
[264] p. : ill., photos. (chiefly col.) ; cm.
Includes bibliographical references.
Summary: Explains the biological processes, the different causes of death, and the human perceptions of death throughout history and across cultures.
ISBN-13: 978-1-55407-996-4 (pbk.)
1. Death (Biology). I. Gingras, Denis. II. Title.
571.939 dc23 QH530.B4558 2012

Library and Archives Canada Cataloguing in Publication

Béliveau, Richard, 1953-
Death : the scientific facts to help us understand it better / Richard Béliveau, Denis Gingras.
Originally published in French under title: La mort.
Includes bibliographical references.
ISBN 978-1-55407-996-4
1. Death (Biology). 2. Life (Biology). 3. Death—Quotations, maxims, etc. 4. Death—Pictorial works. I. Gingras, Denis, 1965- II. Title.
QH530.B4413 2012 571.9'39 C2012-900778-1

Published in the United States by
Firefly Books (U.S.) Inc.
P.O. Box 1338, Ellicott Station
Buffalo, New York 14205

Published in Canada by
Firefly Books Ltd.
66 Leek Crescent
Richmond Hill, Ontario L4B 1H1

Translated from French (Canada) by Barbara Sandilands

Printed in China

To all those who in death taught us more than in life...

Our sincerest thanks to those whose scientific or medical expertise, constructive criticism and humanist vision were instrumental in helping us make this a better book:

Dr. Agathe Blanchette, MD, general practitioner, LMCC, specialized in palliative care and pain treatment, CSSS Trois-Rivières;

Dr. Michel Bojanowski, MD, FRCS(C), professor, Département de neurochirurgie, Université de Montréal, neurosurgeon, Centre hospitalier de l'Université de Montréal;

Dr. Vincent Castellucci, PhD, deputy assistant dean of research, Faculté de médecine, and professor of physiology, Université de Montréal;

Dr. Pierre Dargis, PhD, doctor of physics and physics teacher at Cégep-Limoilou;

Dr. Marie-Claude Delisle, MD, PhD, psychiatrist at the Hôpital Louis-H.-Lafontaine and professor at the Université de Montréal;

Dr. Jean Desaulniers, MD, general practitioner, regional head of continuing professional development for the Association des médecins omnipraticiens de la Mauricie, a branch of the Féderation des médecins omnipraticiens du Québec;

Dr. Pierre Marsolais, MD, internist-intensivist at the Hôpital du Sacré-Coeur, advisor on teaching and hospital development for Québec-Transplant, clinical assistant professor at the Université de Montréal;

Dr. Sergio Faria, MD, radiation oncologist, Montreal General Hospital, McGill University Health Centre;

Dr. Lucie Lessard, MD, FRCS(C), Department of Plastic and Reconstructive Surgery, McGill University;

Mr. Yves Béliveau, playful scholar and inveterate aesthete.

Lastly, our thanks to all those patients who, over the years and in the face of death, have communicated to us their love of life, their fears about death or their end-of-life serenity. Your thoughts, your wisdom and your humor have been a source of inspiration. This book was written because of you.

Foreword

Living is an exciting and rewarding experience. Even though every human life has its generous share of hardships and sorrow, life is above all an opportunity to broaden our horizons and knowledge, take on challenges and fulfill our goals and dreams, be they emotional, professional or material. We are privileged to live in an era when progress in medicine has enabled us to envisage an extraordinary quality of life and a life expectancy never before experienced in human history. As we have said in our previous books, it is even possible to make the most of this longevity by adopting certain lifestyle habits that considerably reduce the occurrence of a number of incapacitating chronic diseases (cancer, cardiovascular diseases, type 2 diabetes, Alzheimer's). A preventive approach of this kind, combined with the enormous curative potential of modern medicine, makes it possible to improve both quality of life and life expectancy, and it thus offers the exceptional opportunity to savor every second of our lives and to participate in the evolution of the society in which we live.

Humans are the only living beings for whom life does not just mean carrying out basic functions for survival and reproduction of the species; it goes without saying that this love of life—as well as the ideals of success and progress we associate with life—make the inevitability of death extremely hard to accept. In an era of overconsumption, in which success is associated much more with acquiring material goods and power

< Gustav Klimt, *Death and Life* (detail)

than with serious reflection on the precarious nature of our lives, death is the ultimate tragic event that we would rather simply ignore, run away from and even deny.

Why write a book on death? Cancer researchers are constantly faced with death. The objective of cancer research is to develop treatments that selectively kill cancer cells while sparing healthy cells. To understand life, it is therefore necessary to understand death and to walk the fine line between the two every day. Thus, our research in neuro-oncology and neurosurgery has enabled us to develop drugs to treat brain tumors, one of the most frightening cancers because it violates the very integrity of what identifies us as a species and defines us as an individual. But, even more importantly, our thinking about death has evolved as a result of the special contact we have had with the seriously ill patients we have been fortunate enough to come into contact with over the years. The depth of their distress or the serenity of their attitude in the face of death have always been an extraordinary source for meditation on the meaning and fragile nature of life. This book is the result of reflection arising from our research as well as from these rewarding encounters.

While it is impossible to prevent death, it is nonetheless possible to assuage the fear of it by better understanding the precarious nature of the processes at work in maintaining life. Science has always played an indispensable role in our understanding of phenomena in the world around us; it can demystify the mechanisms involved in death and cast a fresh eye on what is still the ultimate taboo in our society. To talk about death is to come to grips with the ordeal that awaits us all; by being aware of its inevitability and by better understanding just what it is, we can enjoy every moment of this precious life more. Understanding death better so as to get more out of life—this is our goal in this book.

> The Taj Mahal, built by Emperor Shah Jahan in memory of his wife Arjumand Bânu Begam, is one of the most beautiful mausoleums in the world.

Introduction

> **It is man's destiny to die...**
> **Why should I be sad, when my fate is normal**
> **and my destiny the same as every man's?**
>
> Liezi, *Classic of the Perfect Emptiness,* c. 400 CE

In a famous exchange between a Zen master and one of his disciples, the latter asked, "Master, how can we triumph over death?" The master immediately replied, "By learning to live a better life." Puzzled, the disciple replied, "But, Master, how can we learn to live a better life?" To which the Master replied enigmatically, "Quite simply by triumphing over death...."

This comical dialogue summarizes well the basic dilemma that has haunted human beings since our species first appeared on earth: how can we find meaning in life while knowing that it will inevitably end in death? The driving force behind philosophical inquiry and the great religions, this existential question has preoccupied the greatest thinkers for thousands of years. The works of Plato, St. Augustine, Dante, Descartes, Nietzsche, Heidegger and Sartre (to name but a few) have come down through the centuries and influenced our approach to life in large part because their reflections on the human condition in the face of death have struck a chord in our own questioning about the precariousness of life.

Asking ourselves about the purpose of our time on earth is perfectly normal. For a rational animal like a human being, who is constantly trying to grasp the meaning of the natural phenomena in the world around us, being born to die seems completely incomprehensible, a futile and illogical process that, even though perfectly natural, remains fundamentally disturbing and frightening. It is often said that death is the only

< Vincent Van Gogh, *Old Man with His Head in His Hands*

common denominator in the lives of all people, the "great reaper" who strikes indiscriminately the wealthiest and the poorest, the genius and the idiot, the international star and the most ordinary person. However real it may be, this equality in the face of death nevertheless feels like very small consolation for the loss of our own life and those of our loved ones! Death is a disturbingly ordinary phenomenon in terms of the human species as a whole, with approximately 100 billion human beings having died since Homo sapiens appeared on earth 200,000 years ago. Yet each of these deaths, no matter how anonymous, seemed like a tragic event because it put an end to a unique life, as precious in the eyes of those who died and their loved ones as our own would be today. Death may very well be common and "in the order of things," but it remains nonetheless the greatest ordeal that each of us has to face alone, a predetermined conclusion to a play whose script is being written every day and whose successive events we are constantly trying to stage in a way that makes sense. Before the final act ends and the curtain falls forever, we would really like to understand the reasons for this ending, to attribute meaning to the sudden conclusion of such a wonderful story. To give meaning to life, it is often important to find meaning in death.

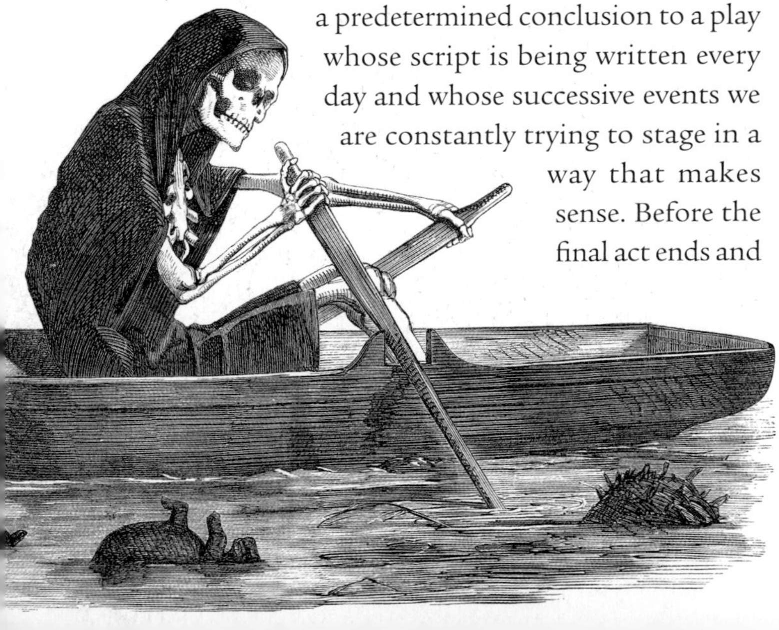

While the history of humanity is marked by stages that have all contributed to the evolution of our species (the discovery of tools, the mastery of fire, the development of language), there is general agreement that the appearance of the first funeral rites is the best indicator of the emergence of modern man (Homo sapiens). Not only do these rites, already evident in the graves of cavemen (around 100,000 years ago), bear witness to the suffering caused by death, they are also the first signs of an attempt to explain its meaning. In this sense, it is striking that, right from the beginning, a number of these graves incorporated elements to facilitate the rebirth of the dead: placing the body in the fetal position to mimic the fertility of the womb; dyeing the corpse with red ocher, likely as a symbol of blood; including everyday objects (e.g., pottery, weapons) to ensure success in the new life. To the question "How can we triumph over death?" human beings' first reflex seems, therefore, to have been to hope that human life is not limited to a brief period on earth, a hope expressed throughout history by the development of more and more complex rituals and religious symbols. And even though these religions have changed considerably over time, they all try to respond to the fear of death by sending basically the same message: earthly life is only a stage, the visible portion of a much longer process that includes rebirth after death.

Nonetheless, we are always taken by surprise when one of our loved ones dies, in whatever circumstances. The death of an elderly person, parent, grandparent, aunt, uncle or other key figures from our childhood is an immensely sad event, even when these people have enjoyed a long life and we can admit that their death is in the order of things. The death of our friends, spouse or colleagues still in the prime of life is a shock that is hard to accept, an ordeal that we go through with a feeling of outrage at the injustice of life. Lastly, and worst of all, the death of a child with their whole life ahead of them is an unnatural event impossible to understand and accept, leaving an open wound than can never really heal. If "neither the sun nor death can be looked at steadily," as La Rochefoucauld wrote in the 17th century, it is precisely because we know it is a constant threat, a terrible and cruel event that takes our dear ones away forever.

But while grieving for and cherishing the memory of our dead is without doubt the noblest expression of our humanity, anxiety about our own death is a heavy burden that can poison our existence. The truth is that a large part of the

^ Baldung Grien, *The Three Ages of Man and Death*

fear caused by death stems from fear of our own death. Whether or not we believe in a life post mortem, death too often remains a taboo subject we prefer to avoid or at least approach with great reluctance, a bit as though, as Freud said, we were unconsciously refusing to accept the idea of our own death. In our opinion, this unease is largely due to a lack of understanding of what death is: Why do we die? What happens when our time comes? Paradoxically, even though all religions and philosophical movements reflect deeply on death's psychological, social and metaphysical aspects, most of us do not know very much about the process of life itself and the events that cause death. We do not really grasp the extent to which a human life is an experience that is as incredible as it is improbable, an absolutely astounding event with its origins in a tiny primitive cell that appeared more than 3 billion years ago. Furthermore, we are unaware that death, far from being a bad or unfair ending to life, has played an essential role in the evolution that resulted in our species' appearance on earth. This is a shame because—paradoxical though it may seem—understanding death helps us to better understand life and to appreciate fully the brief moment of eternity that we have the immense privilege of experiencing, no matter how fragile and fleeting it may be.

With this in mind, we came up with the idea of sketching an outline of what life is and illustrating, using concrete examples, the many ways we can die. Why is cancer such a deadly disease? How can some viruses or bacteria weighing less than a billionth of a gram kill a human being in a few days or even in just a few hours? Why do some wounds cause death while others, which look just as serious, cause only superficial damage? How do you die from poisoning? And even if we manage to avoid all of these ordeals, why do we grow old and ultimately die? We hope to succeed in expressing our belief that understanding the processes that lead to death helps us to better appreciate the limits that are an inevitable part of life and to understand that death really is the only logical ending to existence. Isn't coming to terms with death the best way of getting the most out of life?

Chapter 1
A Material Soul

> The deep sorrow we feel on the death of a friend springs from the feeling that in every individual there is a something which we cannot define, which is his alone and therefore irreparable.
>
> Arthur Schopenhauer (1788–1860)

Some people are terrified of death and prefer to avoid talking or even thinking about it. For others, it is not so much the end of existence that is a source of anxiety, but rather the process of dying and, in particular, the physical and psychological suffering that may precede our final breath. Death is a serious subject that no one is indifferent to, and regardless of our attitude with respect to the end of our existence, the prospect of dying is far from pleasant and always leaves us feeling somewhat helpless.

Even though anxiety about death is inevitable for anyone who loves life, it is possible to alleviate these fears and find a certain degree of comfort by understanding the events that occur in the final moments of our lives. One of the finest qualities of human beings is their insatiable desire to understand the significance of the phenomena in the world around us. This innate curiosity has been a driving force in the amazing accumulation of knowledge that has completely redefined the place we occupy on our planet and shaped the world we live in today. From a scientific point of view, the significance of this knowledge is well illustrated by the many technological advances that have become part of our daily lives, as well as by the extraordinary increase in our longevity, itself a direct consequence of the many advances made by modern medicine. However, the usefulness of the sciences is not limited to the discovery of new technological processes or revolutionary treatments; the sciences must also play a role in changing our

< Funeral mask of Dante Alighieri

way of thinking and our perception of the world by, for example, helping us better understand the factors responsible for our presence on earth and our death. Death is not as mysterious as we often think; on the contrary, it is a perfectly normal and even fascinating event that must be understood better if we are to broaden our horizons and approach life from a new perspective.

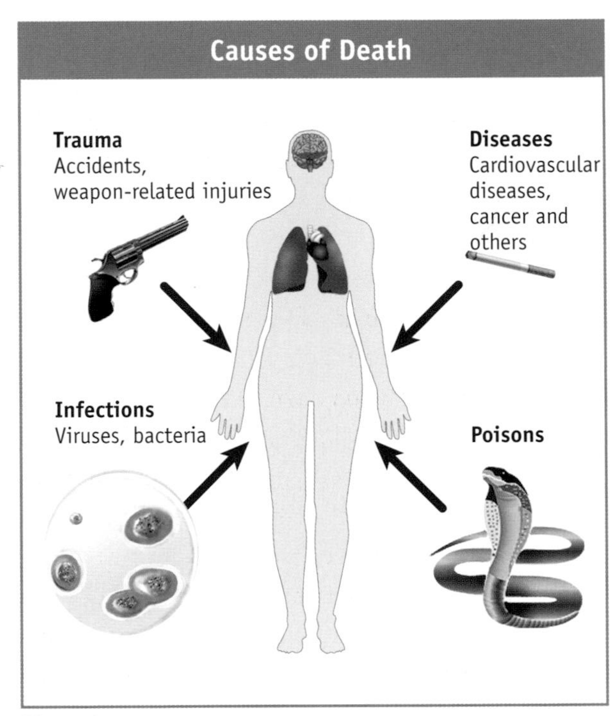

Figure 1

The final breath

The entire spectrum of factors causing death can be divided into just four major categories: (1) deaths due to various illnesses (cancer, cardiovascular diseases, diabetes and genetic diseases, among others), (2) deaths resulting from infections caused by various viruses, bacteria or protozoa (influenza, tuberculosis, malaria, AIDS), (3) those caused by dangerous acts (trauma, death caused by a bullet or bladed weapon) and (4) those resulting from attacks on the organism by various poisons (see Figure 1).

Obviously, these factors have completely different effects on the human body. Hence, as we will see in greater detail in the following chapters, death caused by a virus and death caused by a serious car accident or a sudden cancer involve totally different mechanisms. Nevertheless, in spite of these differences, all these causes of death ultimately stop vital functions in the same way: by blocking, in one way or another, the oxygen supply to the various organs in the human body. Consequently, even though each life is unique and the circumstances surrounding its conclusion are equally so, death remains a relatively simple phenomenon, biologically speaking. Whether caused by disease, infection, trauma or poison, it is always the result of the physiological cessation of the organism's vital functions owing to a lack of oxygen.

Heart or brain?

For thousands of years, the most reliable indicator of a person's death has been the absence of a heartbeat and respiration. For doctors in wartime, for example, a quick and simple method of determining whether a soldier had died was to hold a mirror in front of his mouth to see if the mirror misted up or not.

Even today, death is a state that is sometimes extremely difficult to define from a medical point of view. Since the functions of the heart and brain are so interconnected, it is often hard to determine which of the two is the primary cause of death. Just as the heart sends oxygen-rich blood and nutrients to the brain to ensure the neurons function, certain regions of the brain emit independent neuronal signals that the cardiac muscle absolutely must receive in order to contract.

Because of improvements in a number of resuscitation techniques, particularly the artificial respirator in the 1950s, it can be even more difficult to draw the line between life and death. By keeping the heart and lungs of people in a coma functioning, these resuscitation methods have rewritten the classic definition of death. Indeed, even though the heart is still beating, these methods may result in a state of "brain death," in which the brain completely stops functioning while other vital functions are only

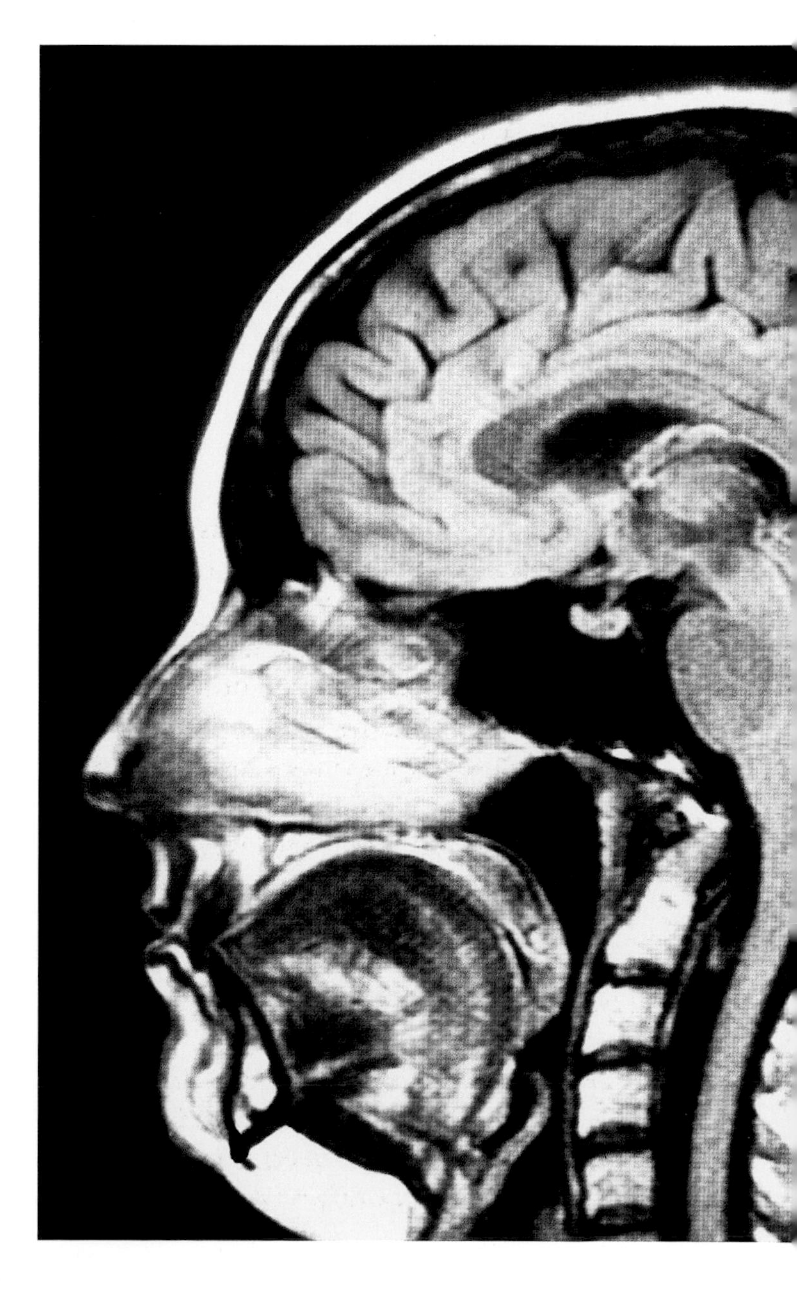

^ Radiograph of a human brain

maintained by artificial means. This situation may be even more complex in that some patients are plunged into a deep coma, yet they can nevertheless survive without any mechanical assistance because the motor centers in the brain that control the heart and lungs remain intact. For example, former Israeli prime minister Ariel Sharon remains in a deep coma more than 6 years after a stroke struck him down in 2006. He is fed by means of a tube and moved around regularly to prevent bedsores, but he will never emerge from this coma, as his brain has atrophied considerably and is only able to keep his vital organs functioning.

Are these people dead? This is a difficult question, the answer to which depends inevitably on how life is perceived. For some groups of people, often the very religious, all forms of life deserve protection, whether embryonic or vegetative. Thus, in the case of Sharon, continuation of care is required by the Jewish faith; according to the Torah, it is forbidden to cause death because of the principle of the absolute sanctity of human life. In mainly secular societies, even though this question sometimes gives rise to heated debates, most people believe that life is more than just maintaining primary physiological functions by artificial means. Death is not solely physical; it is

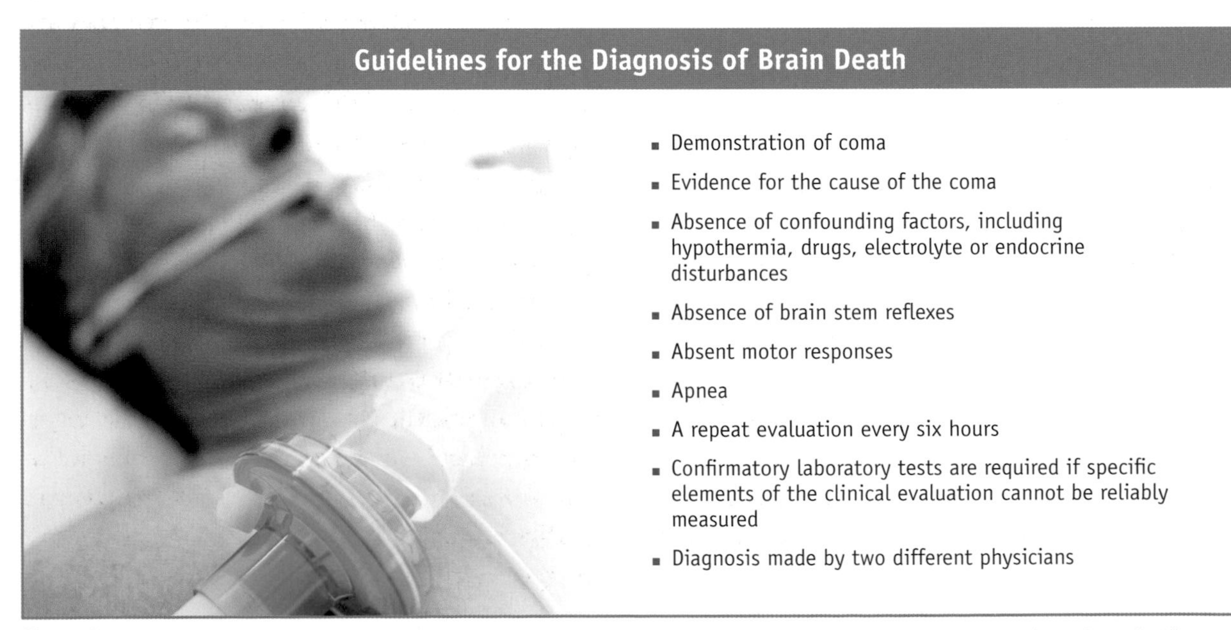

Figure 2

Source: Laureys et al. *Neurology* 70 (2008): e 14–15.

above all the death of a person, of a human being endowed with mental faculties unequaled in the entire living world, which enable him to reflect and interact with his fellow human beings and express his emotions. According to this view, just as our life is different from that of other animals, so is our death. Even though the heart, lungs and organs as a whole are essential to life, it is brain death that marks the frontier between life and death.

Brain death must, of course, be strictly defined from the medical and legal points of view. Currently, a person is declared "brain dead" if and only if his or her neurological state matches the criteria proposed by a group of physicians and bioethicists at Harvard University in 1968. These include, among other criteria, lack of consciousness (coma), lack of reflexes associated with the brain stem (no response to pain, loss of pupillary reflexes, loss of gag and cough reflexes) and apnea (the inability to breathe; see Figure 2). This extremely rigorous diagnosis is considered valid only if the person shows no signs of intoxication, hypothermia or other medical disorders (such as certain serious thyroid-related diseases) because physiological functions in these cases are slowed down to the extreme and can be mistaken for brain death. However, when all these criteria are present and there is no doubt that the person will never regain brain function, the consensus is to declare death.

Evolutionary tinkering

Taking brain death as the ultimate criterion for death is perfectly justifible, since the brain is absolutely essential to life. It acts as a veritable control center, being made up of hundreds of billions of neurons located in distinct regions that collectively coordinate both the maintenance of basic vital functions (respiration, heartbeat, digestion, sexual impulses) and the way we interact with our environment (see Figure 3).

Of course, such a masterpiece of organization did not just appear overnight. It is actually the result of a long evolution over the course of which more complex structures were progressively grafted onto a "primitive" brain, the role of which is to coordinate basic needs. Often referred to as the "reptilian brain," this basic control system corresponds to the brain stem and the cerebellum. It manages the organism's vital functions by controlling, among other functions, heart rate, respiration, body temperature and balance. Unlike that of reptiles, this brain had an "annex" added around 150 million years ago, when a limbic system—a regrouping of structures, notably including the hippocampus, the hypothalamus and the amygdala, which collectively have a determining influence on our emotions and behaviors—emerged in small mammals. However, it is when the cortex emerged that the brain truly aquired great complexity, for the cortex enables the brain to generate advanced functions such as thinking, language, conscience and imagination. The development of the human brain is therefore an excellent illustration of the concept according to which "evolution proceeds like a tinkerer who, during millions and millions of years, slowly modifies his work, constantly retouching it, cutting here, lengthening there, seizing every opportunity to adjust, transform and create" (François Jacob, *Le Jeu des possibles* [The Game of Possibilities], 1981).

Molecular thoughts

The brain owes its ability to generate such abstract phenomena as thought, emotions and intelligence to neurons, a kind of superspecialized cell characterized by the presence of multiple extensions called dendrites (from the Greek for "tree") and axons (see Figure 4). These neurons are excitable cells, which means they can be activated by electrical potential variations and use these variations to transmit information by

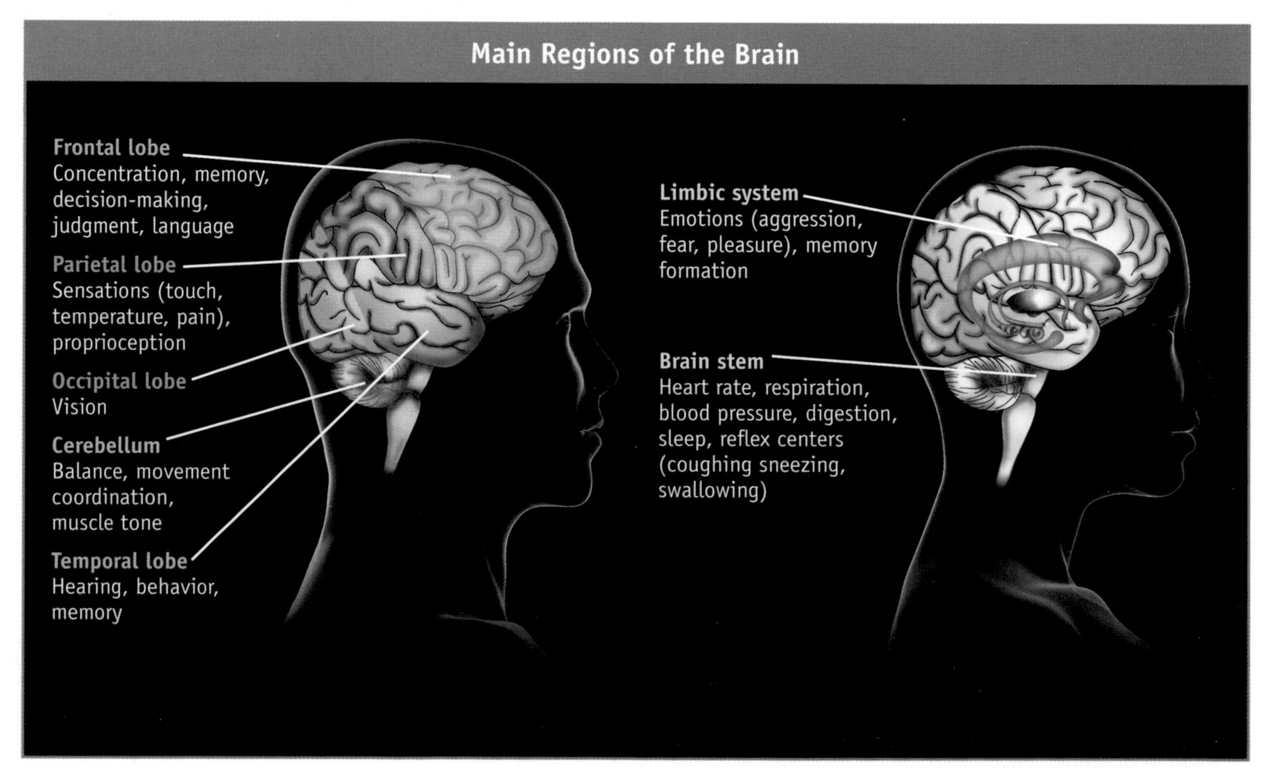

Figure 3

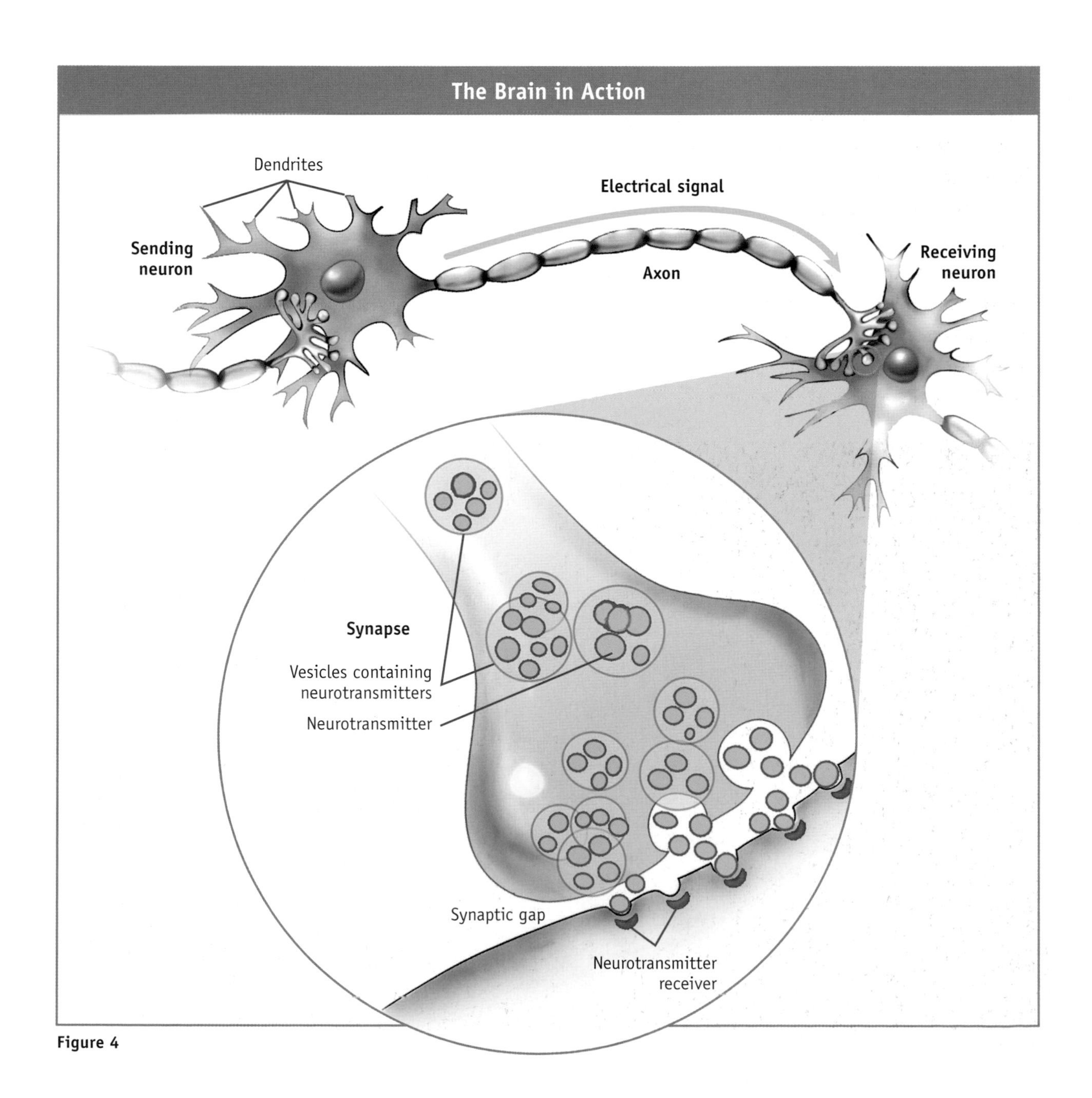
The Brain in Action
Dendrites
Electrical signal
Sending neuron
Axon
Receiving neuron
Synapse
Vesicles containing neurotransmitters
Neurotransmitter
Synaptic gap
Neurotransmitter receiver

Figure 4

connecting themselves to other neurons via connections called synapses. It is estimated that one single neuron establishes, on average, 10,000 synapses through its dendrites and axon. This means that with some 100 billion neurons, the human brain contains about one quadrillion (10^{15}) of these connections. Since thinking is the product of these synaptic connections, it is only normal to reflect on the meaning of life!

Propagating an electric current along a neuron is made possible by the enormous differences in the composition of ions found inside the neuron and that of its external environment; this charge differential creates electrical potential on both sides of the membrane. From an energy point of view, maintaining this potential is extremely costly: even though it makes up only 2% of our weight, the brain alone consumes 20% of our energy (mainly in the form of sugar), of which 80% goes exclusively into maintaining this electrical potential. Considering the advantages that having a large brain offers, you have to admit that the cost is perfectly justified! This electrical energy can, in fact, be recorded using an electroencephalogram, which measures the brain's states of consciousness and activity.

The nerve impulse propagated by a neuron makes it possible to stimulate another neuron across a synapse by means of molecules called neurotransmitters (see Figure 4). When the current running through the axon reaches the synaptic junction, it triggers the release of neurotransmitters into the synaptic gap found at the tip of the axon. These molecules are then able to reach the neuron's dendrites, located nearby (around 40 nanometers away, or a few millionths of a millimeter), where they bind with specific receptors. When there are enough neurotransmitters attached to the dendrites (sometimes it takes the simultaneous firing of 20 synapses to establish a good connection), the signal is transmitted

< Auguste Rodin's famous *Thinker*

> Artistic image of a neuron network

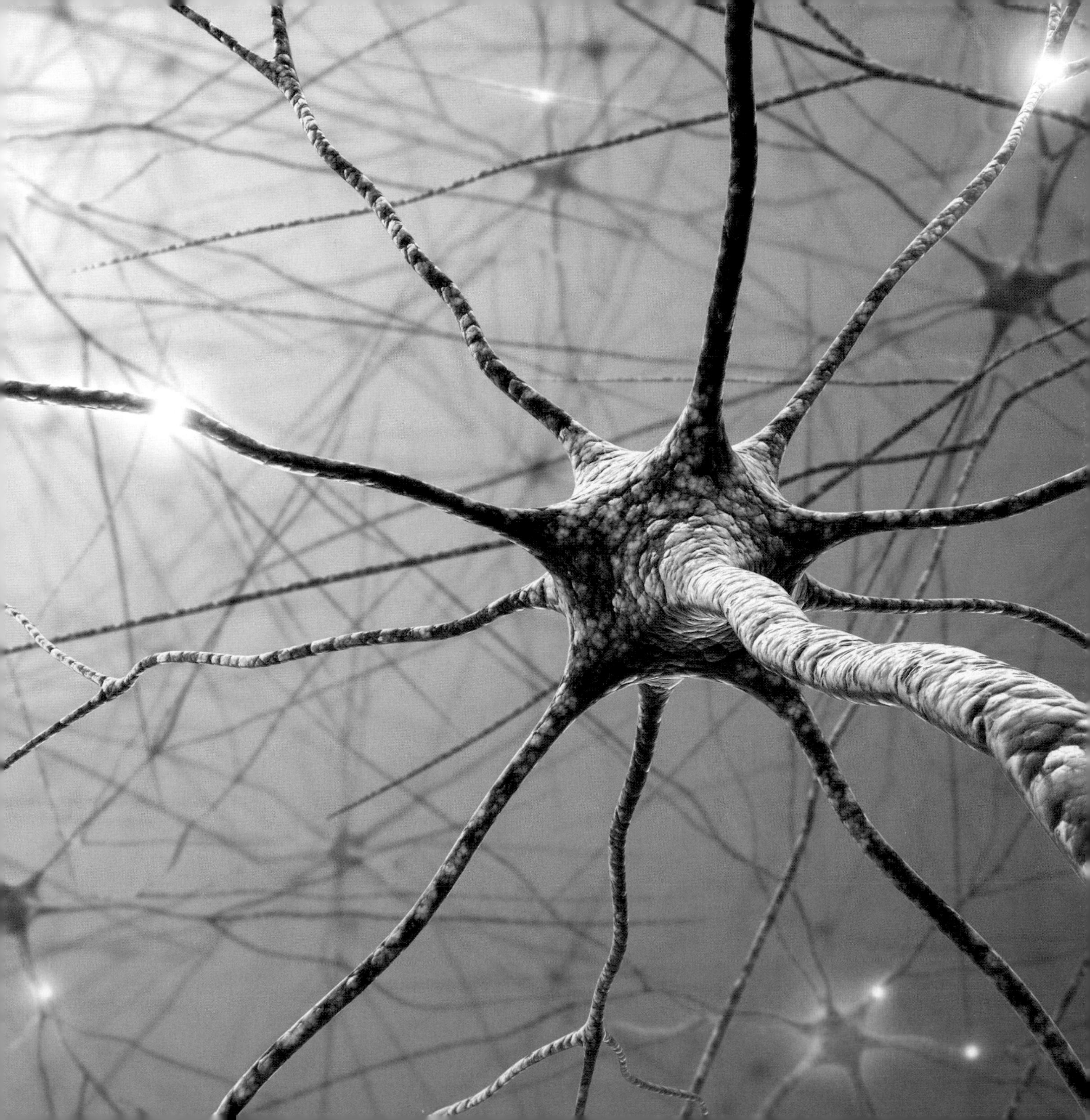

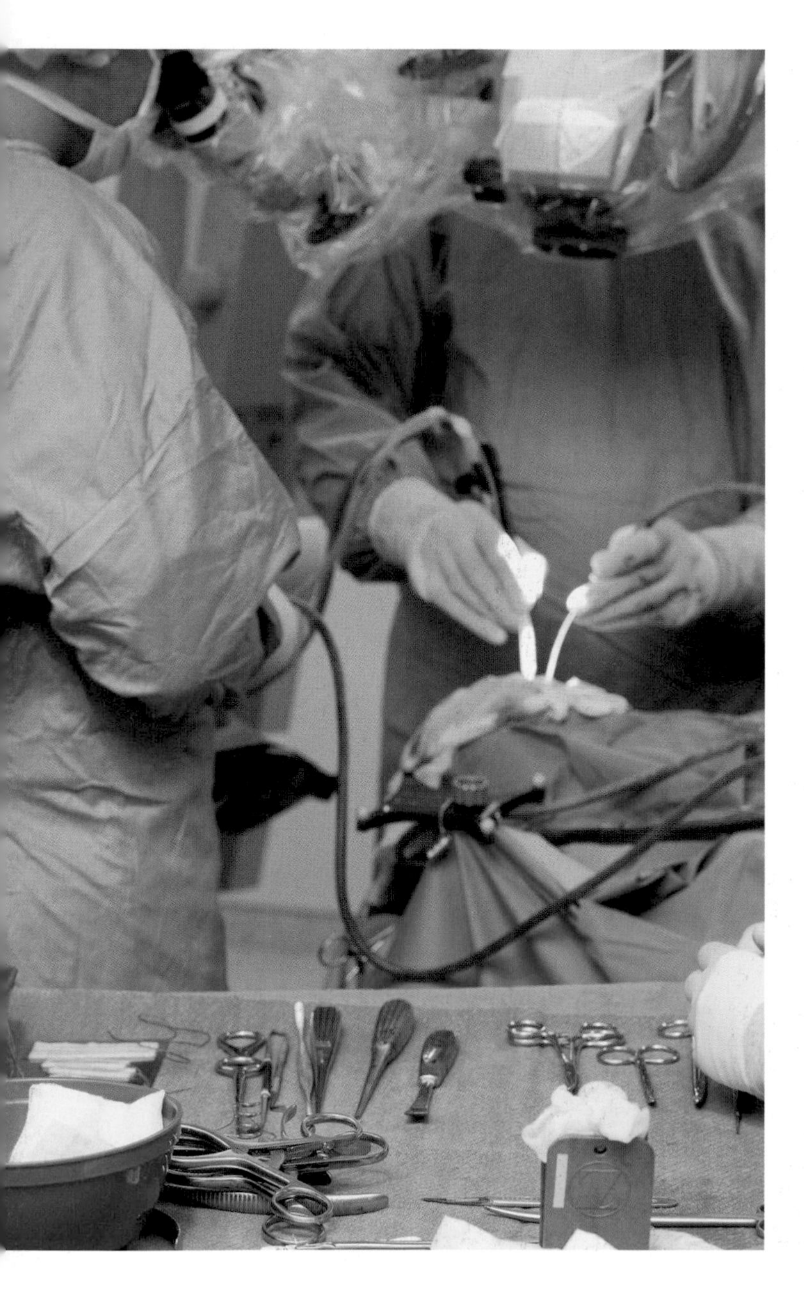

to the receiving neuron and continues on its way. The neurotransmitters that were unable to reach the dendrites of the neuron or bind with them are either recaptured by the sending neuron, using specific transport systems, or destroyed by enzymes occurring in the synaptic space, notably monoamine oxidases (MAO) and acetylcholinesterase. The significance of these enzymes is shown by the rapid death that follows when they are inhibited by certain poisons (see Chapter 7).

The result of a nerve impulse transmission depends on the nature of the neurotransmitter used by the neurons to communicate with each other. Among the approximately 60 molecules that can act as neurotransmitters, some stand out as much for their positive contribution to optimum brain function as for their role as targets for drugs or medications (see Figure 5).

Dopamine, for example, is a neurotransmitter involved in the control of motor functions (the degeneration of a group of neurons producing dopamine is responsible for Parkinson's disease, which is characterized by dysfunctional motor activity, such as trembling at rest and muscular rigidity), as well as in "reward" behaviors, which means it acts on the pleasure centers to encourage the repetition of activities that cause feelings of well-being (eating, sex, drugs, etc.). The sensations of euphoria and happiness associated with alcohol, cocaine, nicotine and amphet-

^ Neurosurgeons performing an operation on a patient's brain.

The Main Neurotransmitters

DOPAMINE is a chemical substance that regulates the stimulation of several regions in the brain and plays a primary role in motivation. When there is a dopamine deficiency, as is the case in Parkinson's disease, poorly functioning neurons make certain movements difficult to carry out. Conversely, too much dopamine may lead to hallucinations and a psychotic state. This is the mechanism that comes into play in cocaine use; cocaine heightens the effect of dopamine while hindering its reabsorption. Nicotine also enhances dopamine activity.

SEROTONIN is a neurotransmitter that plays a role, for example, in moods, anxiety, appetite, sex drive, sleep, pain, blood pressure and the regulation of temperature. A drop in serotonin is linked to some types of depression: at high levels, it makes people feel optimistic and calm. Certain drugs, like Prozac™, Paxil™ and Luvox™, create an antidepressant effect by preventing the recapture of serotonin by nearby neurons.

ACETYLCHOLINE was the first neurotransmitter discovered. It plays a key role in learning, memory and attention span. Acetylcholine deficiency is a factor in Alzheimer's disease.

ADRENALIN is well known as a stimulant. It increases heart rate, raises blood pressure and causes the pupils to dilate. Too much of it makes people anxious.

GLUTAMATE is the brain's main neurotransmitter. It is involved in a third of all synaptic transmissions and plays a role in learning and memory. Glutamate deficiency, therefore, causes difficulties in these two areas.

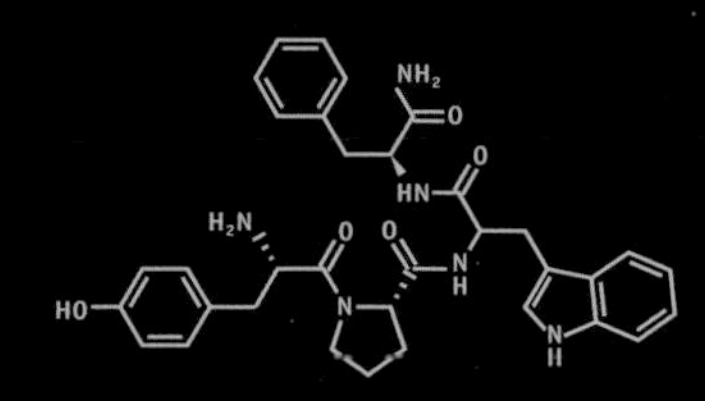

ENDOMORPHIN alleviates pain and causes a feeling of well-being. However, it may be addictive (think of drugs like opium, morphine and heroin). It has also been observed that sugars and fats cause the release of endomorphin.

Figure 5 Source: http://www.linternaute.com/science/biologie/dossiers/06/0602-cerveau/7.shtml

amines are all related, directly or indirectly, to an increase in dopamine in the synaptic junctions.

Serotonin, a molecule related to dopamine and a veritable "happiness molecule," plays a central role in mood control. However, overactivation of the neurons can cause hallucinations, a property long exploited by Mexican shamans; ingesting mushrooms containing psilocybin causes the body to produce psilocin, a molecule that binds to the serotonin receptors, and results in an overstimulation of the serotoninergic pathways, which modifies the perception of the world considerably. The psychedelic effects of LSD are also caused by this molecule binding with these receptors. Conversely, a deficiency of serotonin is associated with a sullen mood and increased risk of depression. Indeed, certain antidepressants, such as Prozac and Paxil, have a beneficial effect because they inhibit serotonin recapture in the synapses, and the subsequent increase in the neurotransmitter's concentration makes the serotoninergic neurons' nervous transmission more efficient.

In addition to their role in controlling thinking, the emotions and behavior in general, neurotransmitters are also responsible for pain perception. From a physiological standpoint, pain is extremely important in keeping the body from being exposed to situations that could damage it (a hand in the fire, for example), as well as for retaining a memory of these situations so as to avoid them in future. The mechanisms involved in pain are extremely complex, but they often call upon the sensory receptors for pain (nociceptors), which are found in various places in the organism. Thus, when these receptors are activated by too high a heat, too much mechanical pressure or an irritating chemical substance, an associated neuron is activated and a signal is transmitted via the spinal cord to the brain to inform it that a danger that might threaten the body's integrity is present. When this threat calls for immediate action, the presence of very rapid involuntary reflexes (the reflex arc) makes an almost instantaneous response to the stimulus possible, even before the nerve signal reaches the brain (pulling the hand out of the fire, for example). At the same time, the brain secretes endorphins, neurotransmitters that stimulate the regions in the brain involved in pain relief and whose role is to alleviate the pain caused by the activation of these neurons. The pain-killing effect of opiates such as morphine and heroin is also related to the activation of these pain-killing neurons (see Chapter 9).

Several mechanisms responsible for physical pain are also at work when a person experiences emotional trauma; this is why pain caused by powerful emotions is felt physically, as if the body was directly affected by impending misfortune. Thus, during particularly difficult

< Michelangelo, *The Creation of Adam,* a fresco on the ceiling of the Sistine Chapel (detail).

emotional experiences, a region of the brain called the anterior cingulate cortex increases the activity of the vagus nerve, a nerve that innervates the chest and abdomen. The overstimulation of this nerve causes the nausea and chest and abdominal pain typical of strong emotions. It goes to show that a painful situation can literally "break our heart" or "make us sick"!

This power of the mind over pain perception can also be illustrated by our attitude when faced with a painful situation. A large part of the pain experienced is subjective and expresses an individual perception, also influenced by one's environment or cultural tradition (see Figure 6). For example, a person who is afraid of needles will feel increased anxiety before receiving an injection; this fear triggers brain activity that mimics pain. This signal magnifies the physical pain of the needle piercing the skin. This effect, called the nocebo effect, is also responsible for the physical aches and pains of hypochondriacs: by repeatedly telling themselves they have a disease, they wind up feeling "real" pain. Conversely, it is sometimes possible to ease a feeling of pain by minimizing its effect, by denying, as it were, that a given event could hurt us. This is known as the placebo effect, a phenomenon in which the brain sets in motion certain processes designed to alleviate the pain somewhat or at least to take our mind off it. The placebo effect explains why we feel better when we learn that the pains in our chest are not caused by a heart attack but by simple gastric troubles. The placebo effect is important in clinical studies where a drug is being tested, as it can account for a significant proportion of the therapeutic response in control groups (up to a third of patients), for whom the drug is replaced by a non-active ingredient. Indian fakirs are an extreme case of the power of the placebo effect. They have developed a heightened ability to block pain perception and are thus able to inflict intense pain on themselves and master it though mental and physical preparation.

Pain Perception

The Placebo and Nocebo Effect

Negative autosuggestion leads to negative effects, known as the nocebo effect. A nocebo effect occurs when thoughts, beliefs or negative expectations are associated with the actual pain and make the symptoms worse.

The placebo effect is the exact opposite. Positive autosuggestion helps alleviate the actual pain.

These two effects show how subjectivity influences experience. One can assist in one's own recovery and help alleviate discomfort, just as one can hinder one's recovery and make discomfort worse.

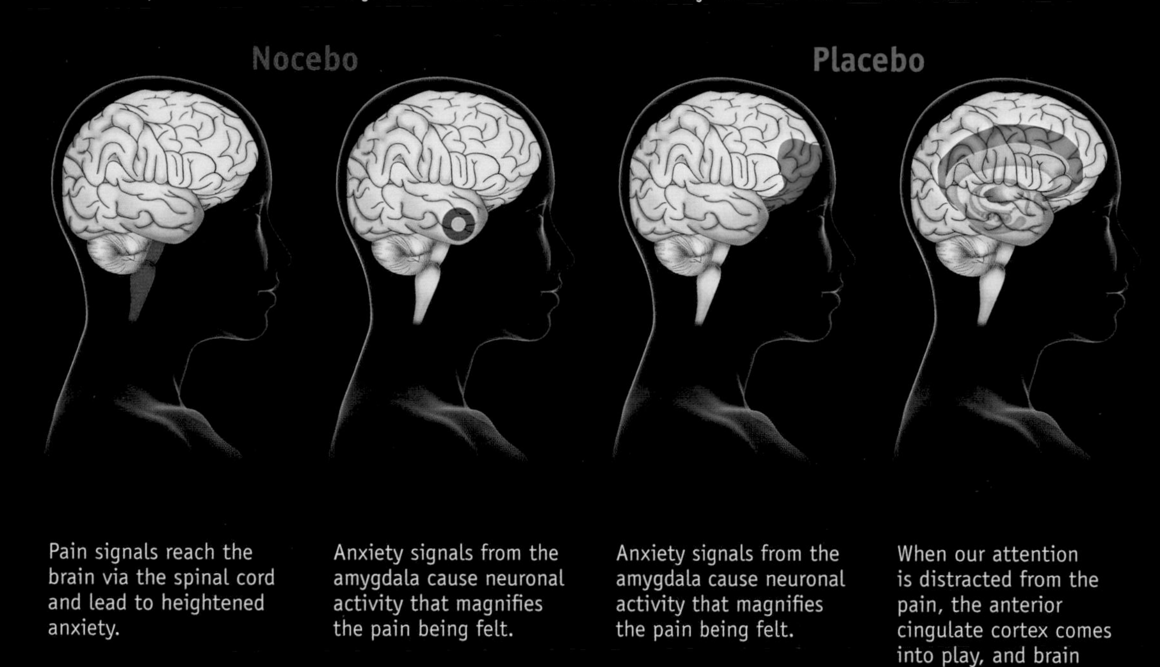

Pain signals reach the brain via the spinal cord and lead to heightened anxiety.

Anxiety signals from the amygdala cause neuronal activity that magnifies the pain being felt.

Anxiety signals from the amygdala cause neuronal activity that magnifies the pain being felt.

When our attention is distracted from the pain, the anterior cingulate cortex comes into play, and brain activity decreases.

Figure 6

< An Indian fakir reclining on a bed of nails.

A cerebral soul

Countless studies based on cutting-edge research in neurosurgery and neurology indicate that the mechanisms involved in nerve impulse transmission in the brain underlie the state of consciousness that defines us as individuals. Brain activity is what enables us to think, owing to the incredible complexity and diversity of the synaptic connections among neurons. It makes it possible for us to reason logically, feel emotions and carry out our many activities as human beings, all expressions of the prodigious capacity of our cerebral cortex, which coordinates these abilities. Conversely, head injuries or serious metabolic disorders resulting from various pathologies (endocrine-related, vascular, hemostatic) will lead to the loss of consciousness that is associated with a comatose state. The seriousness of the coma may vary considerably depending on the damage to the brain tissue (see Figure 7, box on p. 31).

The coma associated with these traumas is caused by a failure in the connections among the various cerebral networks involved in consciousness. Thanks to the development of the neurosciences, especially medical imaging techniques such as positron emission tomography (see box on p. 35), we know that the alert state mainly involves two regions in the cerebral cortex, the temporoparietal and prefrontal regions, as well as the thalamus, which acts as a relay station that transmits and filters sensory information to the cortex (see Figure 8). In an alert state, the precuneus and the posterior cingulate cortex in the temporoparietal region are particularly active, but this activity decreases considerably during general anesthesia or in people in a vegetative state. Certain anesthetic agents deactivate the neurons in the reticular activating system, an area in the brain stem involved in alertness and consciousness, thus causing sleep and coma. Loss of consciousness is thus not caused by the turning off of a central "switch" that controls all brain functions; instead, it results from the disconnection of very specific areas of the brain, whose close collaboration by means of neurotransmitters is absolutely key to maintaining a state of alertness and consciousness.

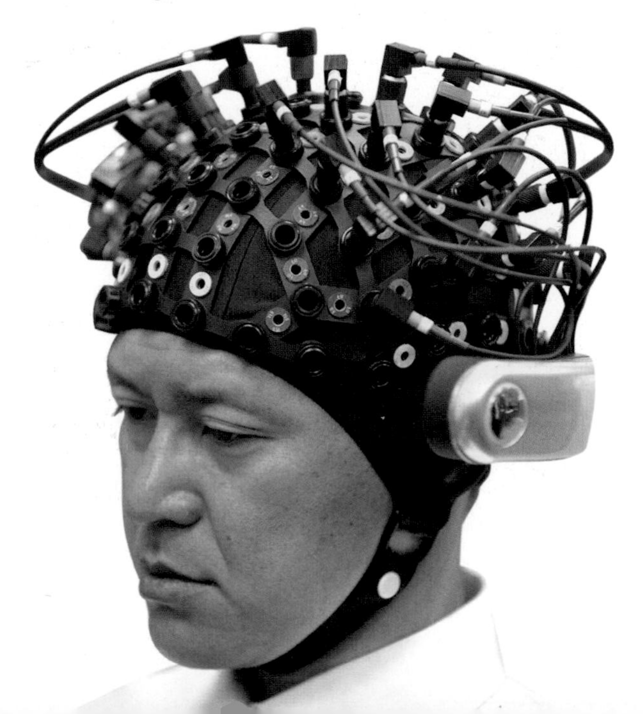

Glasgow Coma Scale

Faculty Observed	Response	Points
Eye opening response	Spontaneous, open with blinking at baseline	4
	Open to verbal command, speech or shout	3
	Open in response to pain applied to the limbs or sternum	2
	None	1
Verbal response	Normal	5
	Confused conversation but able to answer questions	4
	Inappropriate responses, words discernible	3
	Incomprehensible speech	2
	None	1
Motor response	Obeys commands for movement	6
	Responds to pain with purposeful movement	5
	Withdraws from pain stimuli	4
	Responds to pain with abnormal flexion (decorticate posture)	3
	Responds to pain with abnormal extension (decerebrate posture)	2
	None	1

Combined scores < 8 are typically regarded as coma.

Figure 7 Adapted from Teasdale G, Jennett B, "Assessment of Coma and Impaired Consciousness: A Practical Scale." *Lancet* 2:81–84 (1974).

Loss of consciousness

The word "coma" comes from the ancient Greek for "deep sleep." It is a complex state characterized by the absence of verbal, motor and ocular responses to a stimulus. The seriousness of head trauma and coma can be determined using the Glasgow scale, a tool created in 1974 to assess the chances of a patient's recovery. This scale is based on a 15-point evaluation of motor, verbal and ocular functions (see Figure 7).

The total points obtained in each category make it possible to estimate the extent of brain damage. A score of 13–15 is usually associated with minor trauma, whereas a score of 8 or less is typically that of a person in a coma.

Vegetative state In the most serious cases involving major injuries to the brain (Glasgow score 3), the affected person is said to be in a "vegetative state," meaning that there appears to be no superior brain function and that all ability to respond to external stimuli or communicate with others has been lost. Basic functions like breathing, heartbeat, the waking-sleeping cycle and certain reflexes are, however, preserved if the diencephalon (thalamus, hypothalamus) and the brain stem are intact. In such circumstances, patients show certain complex reflexes coordinated by these structures: yawns, chewing move-

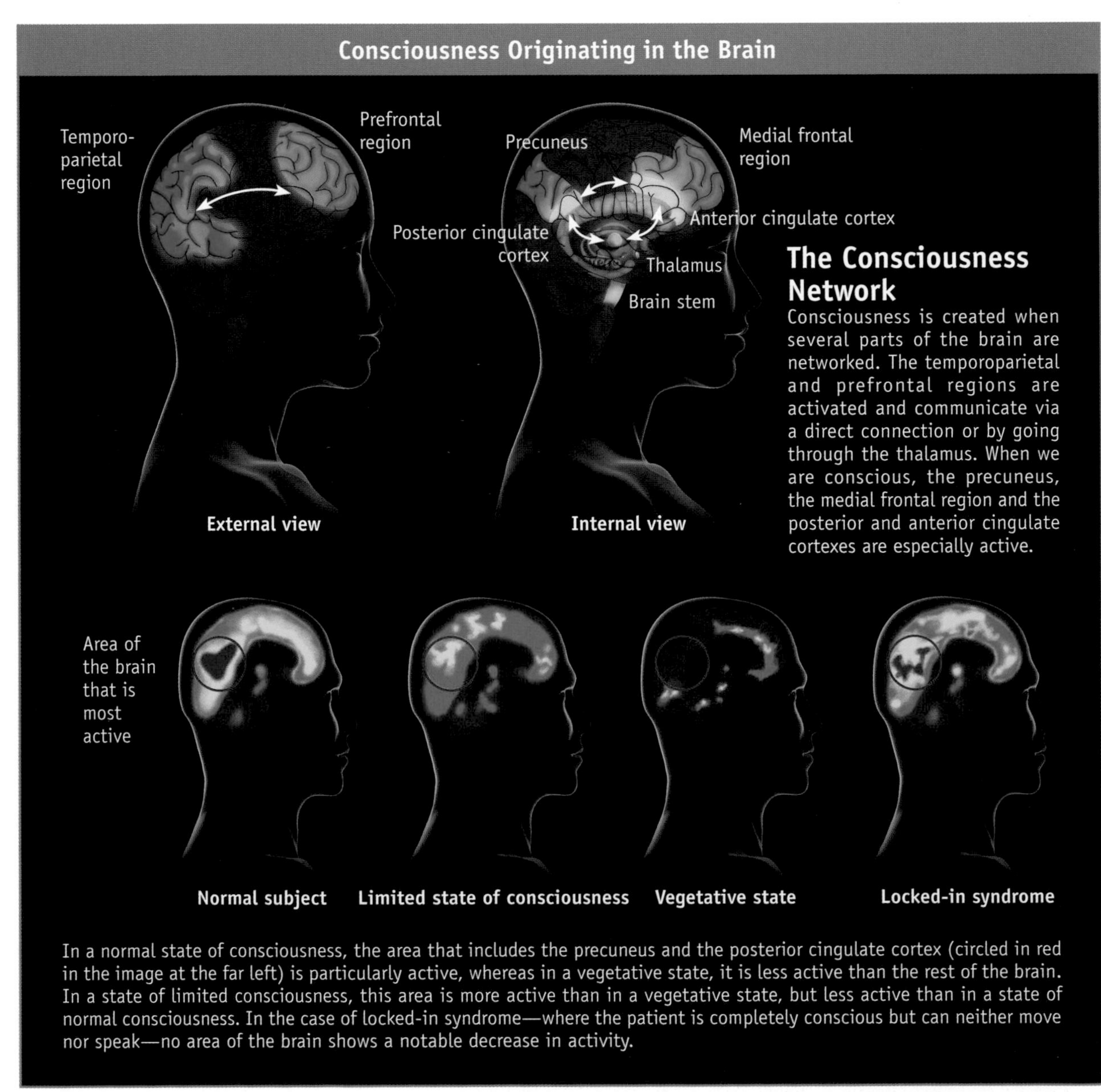

Figure 8

From *La Recherche*, March 2010

ments and sometimes even the emission of guttural sounds, eye movements (reactive pupils, the oculocephalic reflex) or arm and leg movements (a motor reflex that consists of squeezing an object when it comes into contact with the hand). Prolonged arrest of brain circulation caused by strokes or head injuries is often the main cause of a vegetative state. If this state lasts longer than a month, the patient is deemed to be in a permanent vegetative state and the likelihood of recovery decreases considerably. Locked-in syndrome, on the other hand, is a neurological state in which the patient is awake and conscious, and with adequate sensory perception, but is unable to speak and move. It is usually the result of a stroke.

Cerebral death Also called brain death, cerebral death results from the total failure of brain functions, including those controlled by the brain stem. The patient cannot breathe without assistance and is only kept alive using medical devices to sustain basic cardiopulmonary functions. This loss of basic brain function is confirmed by the total absence of brain stem reflexes (no contraction of the pupils to light, no reaction to pain, no cough or gag reflex when a catheter is inserted into the trachea). According to the clinical practices in various countries, a lack of blood circulation to the brain is visualized using cerebral angiography or scintigraphy, whereas the absence of brain activity can be confirmed by a flat-line EEG. When there is this degree of brain inactivity, recovery is absolutely impossible; this is why, in most parts of the world, a diagnosis of brain death is considered to mean the death of the person and laws allow treatments to be stopped and artificial life support to be terminated.

Organ donation

Brain death, which is considered to mean death, occurs despite the fact that the vital organs, such as the heart and lungs, are still intact and functioning well. If the deceased expressed his intention to donate his organs after death, they can be removed and transplanted, as they are still well supplied with blood and oxygen. The extraordinary progress achieved in recent decades in the field of organ transplantation (kidney, heart, lung, liver, pancreas, intestine) and tissue transplantation (corneas, skin, heart valves, among others), both in terms of surgical procedures and the control of immunosuppression mechanisms essential to the survival of the graft, means that the lives of many people waiting for transplants can be saved. By better understanding the issues surrounding brain death, we can, through our generosity and after our own death, prolong another person's life. Even in death we can continue to express one of our species' finest characteristics: compassion, rooted in the solidarity of its individuals.

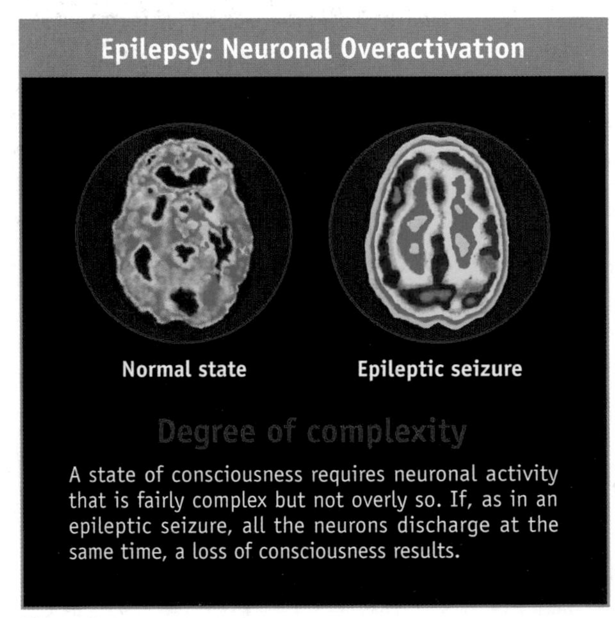

A state of consciousness requires neuronal activity that is fairly complex but not overly so. If, as in an epileptic seizure, all the neurons discharge at the same time, a loss of consciousness results.

Figure 9

On the other hand, the intensity of the signal generated by these communication networks must be tightly controlled: during epileptic seizures, for example, the activation of neurons is much too intense and causes instead a loss of consciousness (see Figure 9).

Molecular moods

Several other observations show that the mechanisms responsible for transmitting nerve impulses in certain very specific areas of the brain are responsible for controlling personality.

Brain traumas Certain head injuries can cause significant behavioral disturbances. The most famous case, which established the connection between the brain and behavior, is undoubtedly that of Phineas Gage, a 25-year-old foreman who, in 1848, was working on building a rail line (the Rutland Railway) near the village of Cavendish, Vermont. While he was using a metal rod to ram explosives into a hole in the rock, an unexpected explosion occurred, propelling the 13-pound (6 kg) rod (3 feet/90 cm in length and 1½ inches/3 cm in diameter) through his left cheek into his head, destroying his brain's left frontal lobe before landing 85 feet (25 m) away (see Figure 10). As incredible as it may seem, Gage regained con-

Reconstruction of the Head Trauma Experienced by Phineas Gage

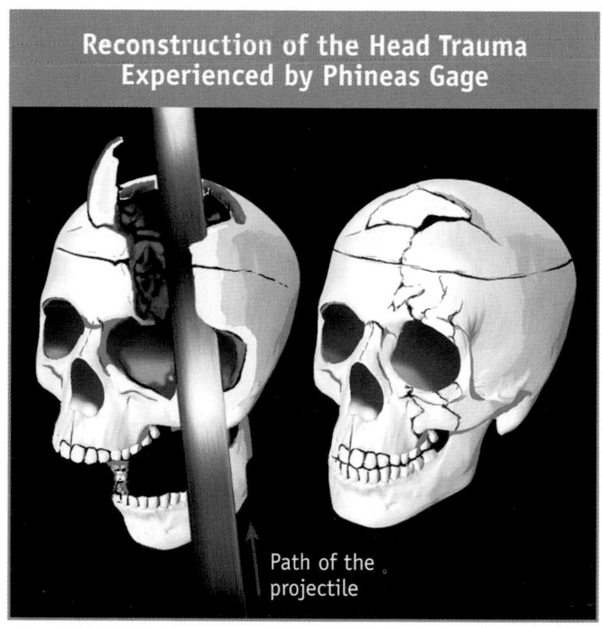

Figure 10 Source: *NEJM* 351 (2004) : e21.

PET and MRI

Brain activity is visualized using positron emission tomography (PET). Using an oxygen-15-water tracer, it is possible to measure blood flow, which reflects the metabolic activity in a given region of the brain. This marker emits positrons that produce photons detected by a camera. This method of analysis is very fast and can be repeated at short intervals, resulting in dynamic imaging of brain function or disruption. Another imaging technique, magnetic resonance imaging (MRI), produces very detailed anatomical images. When these two tools are combined, the anatomical region of the brain where a specific mental activity takes place or where it no longer occurs can be precisely determined.

sciousness some moments later. He had lost the use of his left eye but did not seem to be experiencing any serious physical trauma. However, whereas he had been a model employee and liked by all his colleagues before the accident, his temperament and personality changed dramatically following the trauma. He became rude, temperamental and incapable of forming normal relationships with the people around him. The man who "was no longer Phineas Gage," according to his friends, died 12 years later of an epileptic seizure. Since then, many studies in traumatology and experimental surgery have demonstrated the impact of physical trauma on specific regions of the brain in terms of behavior.

Near-death experiences Many people emerging from a coma have reported experiencing a series of events suggesting that they were on the verge of death, notably the sensation of floating outside their body, an intense feeling of well-being and the impression of entering a bright tunnel with a light at the end of it. These kinds of out-of-body or near-death experiences (NDEs) have been described since time immemorial. They have always held great fascination for human beings, influencing how we perceive the mind and the body, regardless of culture or religious beliefs. From a scientific point of view, these out-of-body experiences have three main characteristics: the feeling of being outside our own body; the impression we are seeing everything around us from an aerial perspective (an extracorporal egocentric perspective); and the impression we are seeing our own body from this perspective, called autoscopy (see Figure 11).

People who have described these phenomena were clearly not dead and their cerebral cortex was still functioning. Numerous neurological studies indicate that these out-of-body experiences seem to be related to the faulty processing of sensory information in the temporoparietal junction, an area of the brain key to self-representation and self-awareness. This is why the electrical stimulation of regions in the temporal lobes of epileptic patients has been shown to be sufficient to induce hallucinations and sensations of body displacement. Even though we still don't precisely know how the molecular mechanisms responsible for

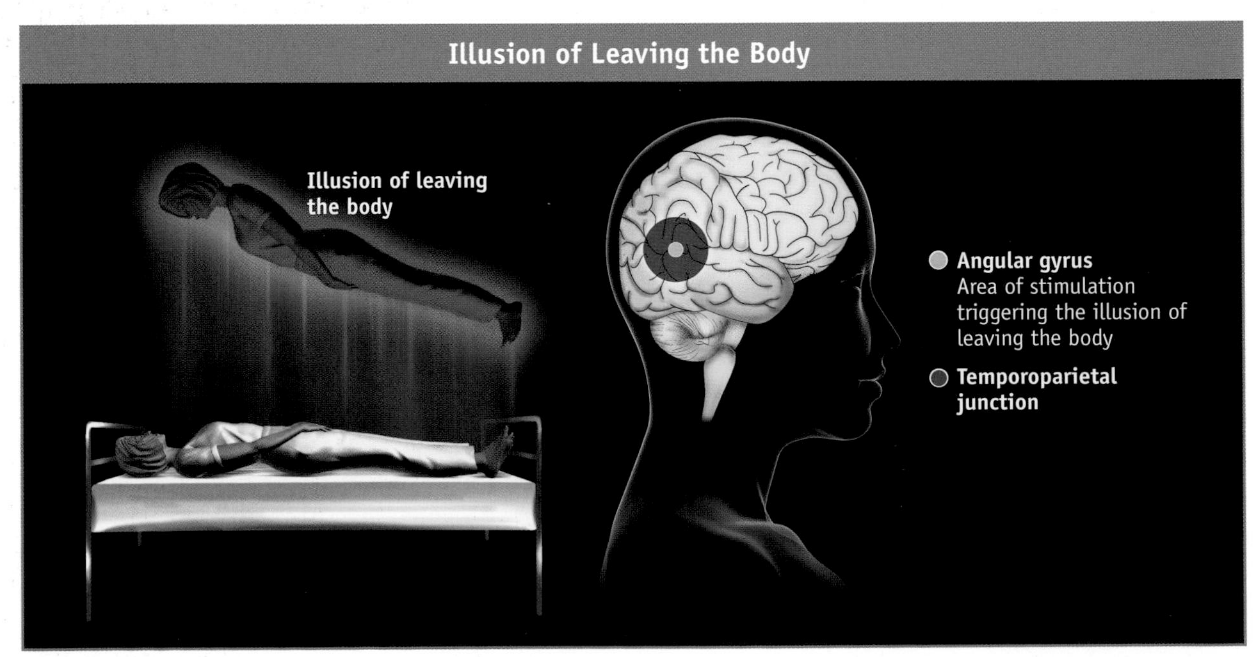

Figure 11

from *La Recherche*, March 2010.

this loss of information processing capacity work, it is nevertheless interesting to note that certain drugs, like ketamine and ibogaine (the active molecule in the strongly hallucinogenic iboga plant), manage to recreate several typical out-of-body sensations. Since the disassociation of the mind from the body caused by ketamine correlates with the interaction of the drug with receptors for the neurotransmitter glutamate, it is likely that certain neurons activated by this neurotransmitter are involved in this phenomenon. It is also known that hypoxic conditions (a low level of blood and oxygen in the brain) as well as an increased level of carbon dioxide (CO_2) are conducive to out-of-body experiences and cause high levels of glutamate secretion. This oxygen deficiency may then alter the message carried by nerves from the visual, auditory and emotional structures (memories and emotions) and cause feelings of well-being, visions of tunnels and the impression that the person's life is passing before his or her eyes often associated with NDEs. Hence, as strange and dramatic as they may be, these out-of-body experiences remain first and foremost illusions and sensations resulting from a disturbance in the brain, the consequence of a traumatic experience in which conditions critical to survival are such that life is threatened.

Genetic variations Even though behavior is the result of a complex interaction between genes and several environmental variables (education, lifestyle, culture), a number of studies nevertheless suggest that genetic variations might exert an influence over various aspects of one's personality. For example, the monoamine oxidase A (MAOA) gene is often considered to be the "warrior gene," a nickname stemming from the following observation: some families with a history of violent criminal behavior over the course of several generations have a mutation that inactivates this enzyme. Since MAOA plays a role in the degradation of the neurotransmitters dopamine, norepinephrine and serotonin in the synaptic junctions, it is possible that the overstimulation of neuronal circuits using these neurotransmitters might cause a behavioral disturbance that, in a particular social context, increases the probability of impulsive aggression.

> An American soldier deployed in southern Iraq in 2003.

Variations in certain genes that participate in the transmission of nerve impulses by serotonin neurons have also been observed, and these genes could increase the risk of disorders as varied as the tendency to become depressed after a trauma and even sudden infant death syndrome. Such variations also exist in the mechanisms that cause pain, the most famous instance being congenital insensitivity to pain. A mutation in the sequence of a pore (sodium channel $Na_V1.7$) found in neurons activated by nociceptors prevents the transmission of nerve impulses and makes people who have this hereditary disease completely insensitive to pain. While this congenital insensitivity to pain is very rare, it is well-known to crime novel fans thanks to Lisbeth Salander's extremely dangerous half-brother, a character in Stieg Larsson's Millennium trilogy.

Body and mind: Infinite variations on the same theme

The relationship between body and mind is a subject that has fascinated human beings for thousands of years. This interest is completely normal, given the central role that intelligence plays in our lives. Exclusive to humans, this ability to think, reason, communicate and express emotions has long fascinated philosophers, who could not conceive that such noble activities could be compared to the "animal" functions of, for example, digestion and muscular movement. This is why, very early in the development of the main human cultures, all activities involving the brain (thinking, feelings) were considered to be non-material phenomena, since the abstract nature of rational thinking could not, according to this view, be caused by actual physical processes. This perception, defended vigorously both by the Greek philosophers such as Plato and Aristotle and by early theologians such as St. Augustine and St. Thomas Aquinas, still exerts an enormous influence on the way we perceive life (and death) on earth. It implies that human beings are different from other animal species because they are composed of two essences, one that is material—the physical and mortal body, and another that is non-material—whether mind, consciousness or soul, existing outside the rigid laws of

< René Descartes

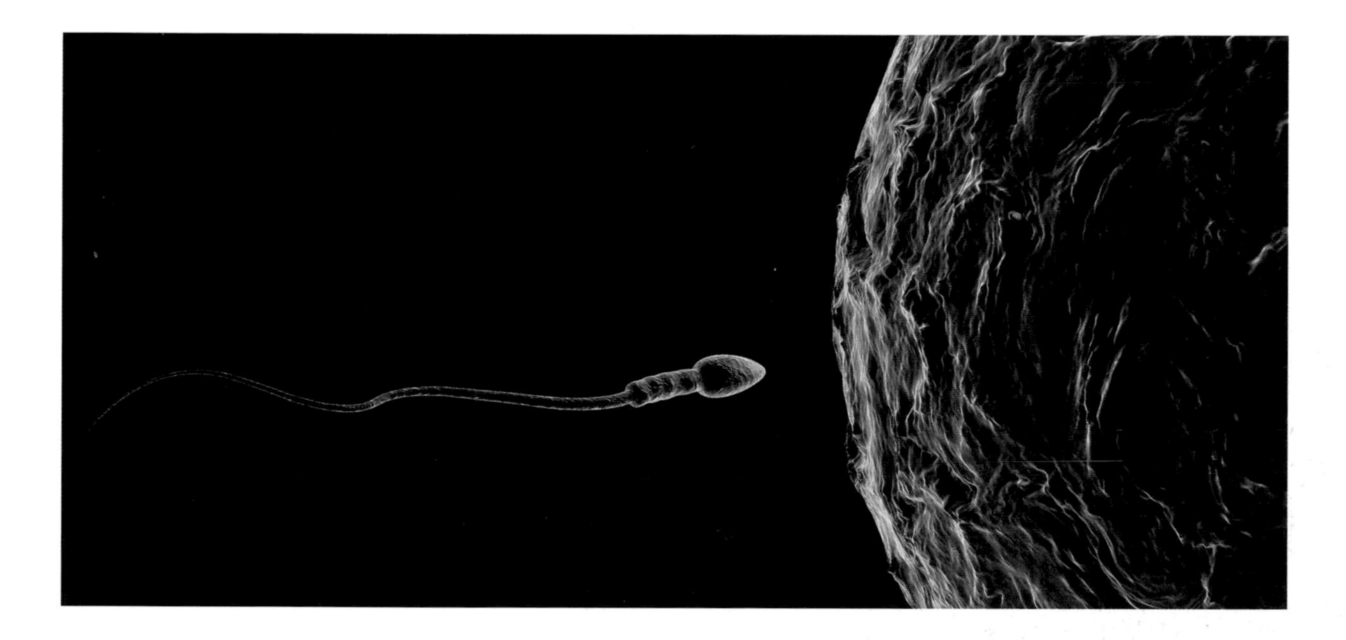

nature because of its immortality and, therefore, somehow able to "survive" the death of the physical body. According to this dualism, our personality—the thoughts, behaviors and emotions unique to each of us—is rooted in inexplicable metaphysical phenomena occurring in parallel with our vital day-to-day functions. Research in genetics and neurobiology, however, has given us some answers to help resolve this dualistic ambiguity and explain more scientifically the complexity of what it means to be human. In reality, the remarkable nature of human life is shown in the richness of individual personalities, resulting from the fact that we are biologically unique. A human being is the product of an astonishing interaction between the genes inherited from one's parents (genetics) and all the modifications that lifestyle imposes on these genes (epigenetics). When the reproductive cells are formed, the 23 pairs of chromosomes that carry our genes are distributed randomly, with one chromosome from each pair either transmitted to an egg cell in the woman or to a sperm cell in the man. From a statistical point of view, this distribution means that 2^{23} (8,388,608) different reproductive cells can be produced. Since, at the moment of fertilization, one of the father's 8,388,608 possible sperm cells randomly fuses with one of the

^ Sperm cell and egg cell

mother's 8,388,608 possible egg cells to produce an embryo, one couple alone can give birth to 70,000 billion different children. In addition to the incredible diversity of their genetic heritage (their heredity), each of these children will live a distinct life, influenced by where he or she grows up, the other people he or she meets, the tastes he or she develops over time and the skills he or she hones through experience. Taken together, these influences have an effect at the molecular level through what are called epigenetic modifications, which do not affect our genetic code but rather the molecular mechanisms that regulate its expression. These profound differences between human beings are not uniquely morphological or anatomical; they affect human behavior in its entirety, including, for example, emotions, fears, artistic sense and athletic abilities. This extraordinary polymorphism is what gives rise to the wide range of human personalities. As the unique nature of fingerprints reminds us (even identical twins show some notable differences in the shape of their dermatoglyphics), every cellular process at work in the human body is a consequence of this unique genetic combination. We are the result of the interaction between a grouping of genes that never existed before and will never exist again and all the modifications that our life imposes on the expression of these genes. This is where the unique character of an individual personality resides.

Like all organs, the brain is not exempt from this remarkable biological diversity. The fact that each individual possesses a unique personality is largely because this genetic and epigenetic diversity creates synaptic connections specific to each of us, as well as individual variations in the relative concentrations of neurotransmitters involved in the transmission of nerve impulses. These neurological phenomena are directly linked to our genes, as well as to the influence that our lifestyle has on how they function. Personality disturbances caused by changes in the biochemical mediators in the brain's nerve impulses in the presence of drugs and medications, by the mutation of certain genes or by serious traumas show the extent to which these mechanisms play an essential role in who we are as individuals.

Thus, the brain is not just a sophisticated regulator for our basic functions; it is also the seat of our thoughts, our memories and our emotions, the chief executive officer in charge of our personality and our identity. What we call "death" is therefore the irreversible loss of the brain functions that make us unique, whether this loss is the cause of cardiac arrest or the consequence of it. Death is really the death of this cerebral soul, the signature of our identity.

> Rembrandt, *The Anatomy Lesson of Doctor Tulp* (detail)

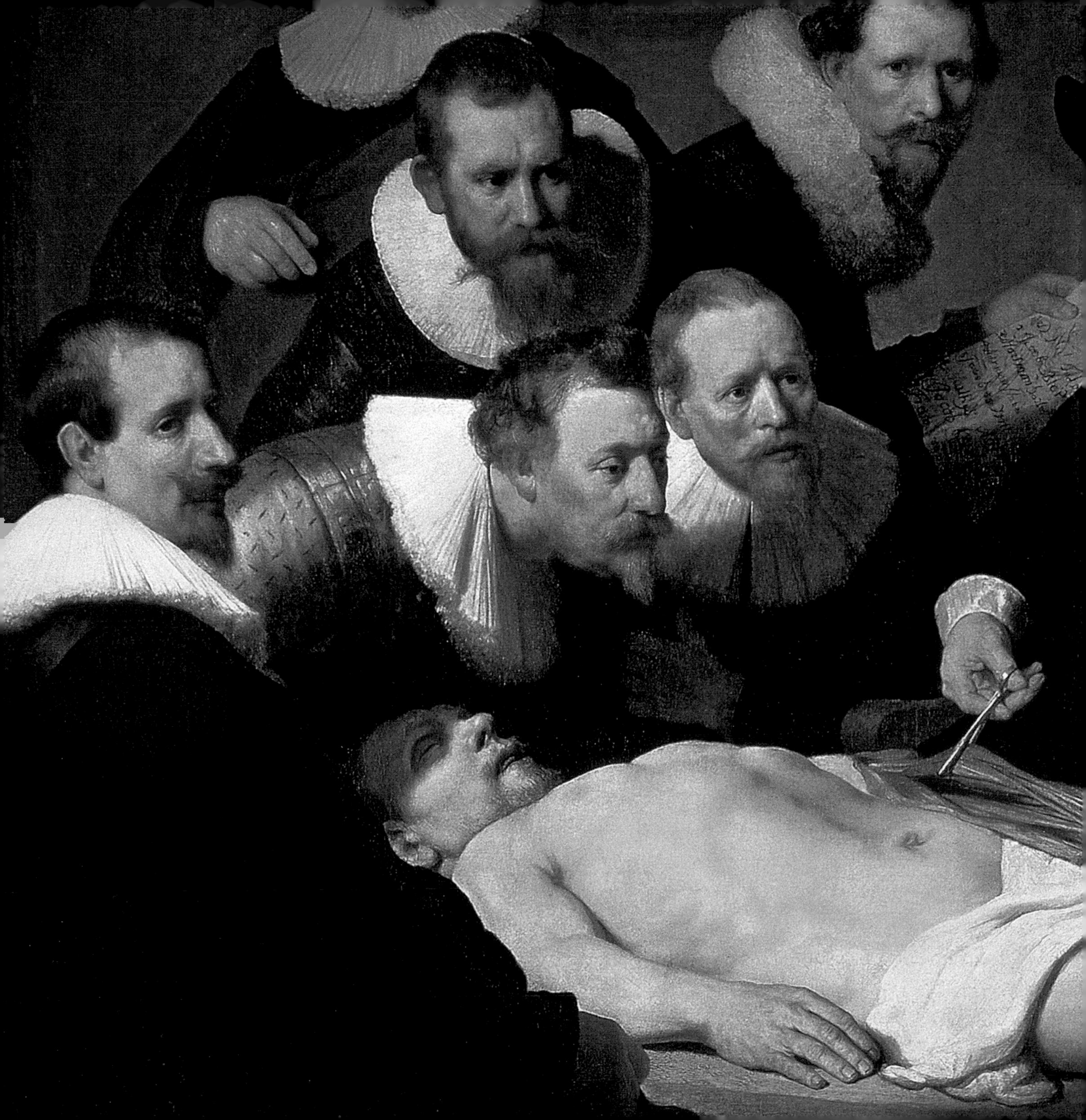

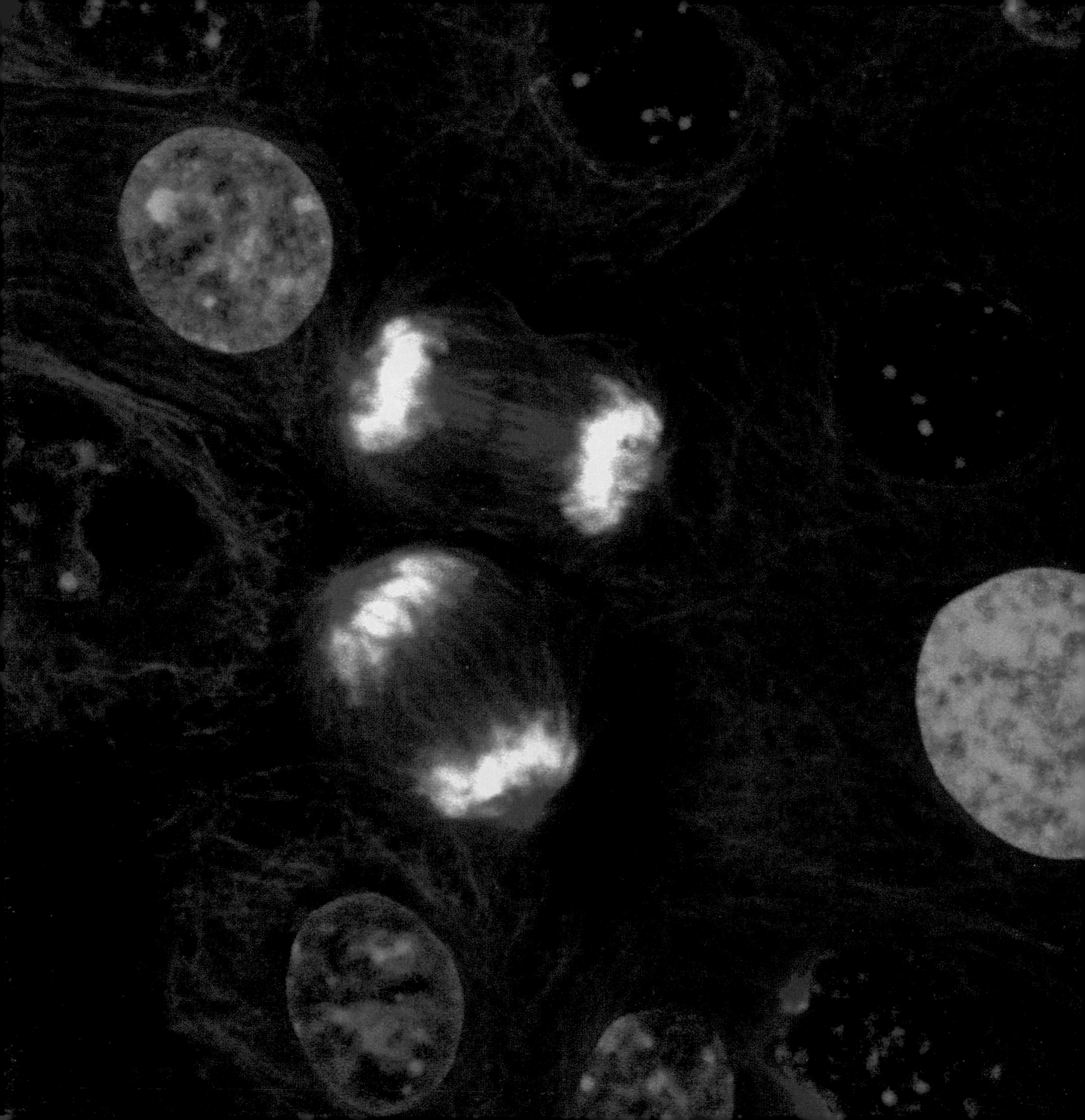

Chapter 2

Dying—That's Life!

> If we don't know what life is,
> how can we know what death is?
>
> Confucius (551–479 BCE)

It is difficult to understand and accept death without first appreciating the complexity of life. We must be aware that our existence, like that of all living organisms on our planet, is the extraordinary result of the evolution of a tiny primitive cell that appeared nearly 4 billion years ago. The conditions required for life to blossom are so difficult to create that we have not yet found any trace of life-forms on any other planet in the explored universe. Given life's improbability and incredible complexity, before asking ourselves why disease occurs or why life one day ends in death, we must first marvel at the fact that life managed to burst forth on earth, resulting in the tremendous variety of species, including our own of course, that have lived and still live on our planet.

Life is beautiful

Life is a constant source of wonder. How can we not be astounded by the incredible work done by neurons, as a result of which we can think and retain memories of important events? By the ability of an immune cell to recognize and literally swallow a pathogenic bacterium to protect us against external attacks? By the mechanism that enables retinal cells to capture the photons in light, making it possible for us to see and appreciate the beauty of the world around us? How can the fusion of a single egg cell with a single sperm cell create a complex human being, composed of 100,000 billion specialized cells that work together to embody this incredible

< Electron microscope image of cell division

experience we call life? Although we are dazzled by technological progress or lured to new gadgets that are constantly being advertised, most of the time we are not aware of the perfection of the cells making up our body. Our daily activities, completely unremarkable on the surface, such as brushing our teeth, threading a needle or hammering a nail, nevertheless require an incredible number of nerve signals to coordinate the visual signal, the exact position of the limbs and the intensity of muscle contractions. Unfortunately, it is often only when we get old or become ill that we come to truly appreciate the quality of life we enjoy when our organs are functioning well and that we understand what being in good health means.

The evolution of life

Becoming familiar with our family tree and learning the names and broad outlines of our ancestors' lives is a concrete way of finding out more about the people to whom we owe our existence.

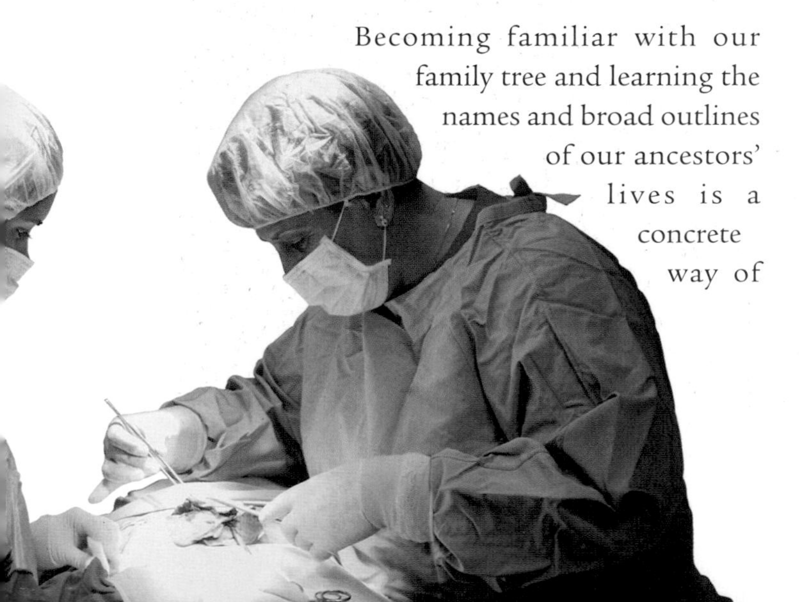

However, it is often difficult to learn with any accuracy the identity of our ancestors further back than the 15th generation (about 400 years), since older documents were unfortunately very often destroyed or lost in the wake of the many imponderables of history. We face the same problem in trying to trace the genealogical tree of life on earth. In fact, although some early forms of life have left traces we can see, in the form of fossils dating back several million years, these fossils only form under very particular conditions and, therefore, represent just a tiny portion of the species that have lived on the planet. Luckily, the enormous progress made in studying the genetic material of a large number of currently existing species has made it possible to identify their similarities and thus determine the degree to which they are related and their common ancestors. Thanks to what is really "molecular genealogy," it is possible to go back in time and trace the general outlines of the stages that led to the appearance of the species that inhabit the planet today. According to data currently available, the three main groups of living organisms—bacteria, archaea (organisms similar to bacteria, but which often live in extreme conditions) and eukaryotes—are all distinct families with a universal common ancestor (LUCA), which appeared

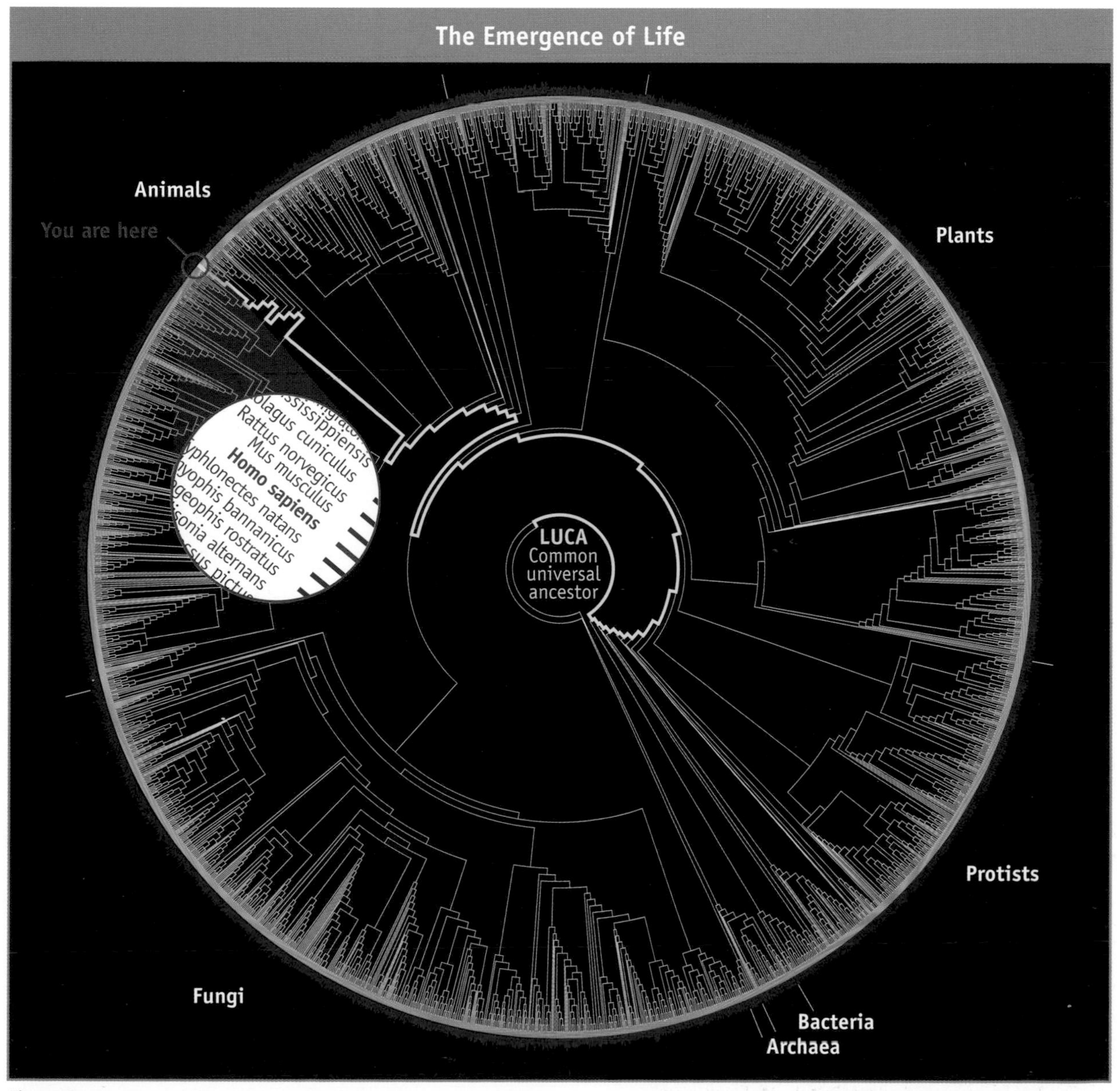

Figure 1

Source: David M. Hillis, Derrick Zwickl and Robin Gutell, University of Texas.

The emergence of life

Even though the possibility that forms of life could have emerged on another planet in our universe cannot be excluded, it is nevertheless certain that life as we know it on earth is an extremely rare phenomenon. The circumstances under which life appeared nearly 4 billion years ago are becoming increasingly better understood. In 1953, Stanley Miller, a chemist, was the first to show that the extreme atmospheric conditions prevailing at that time on our planet (the presence of methane, hydrogen and ammonia gases combined with intense electrical activity) were able to spontaneously create certain basic elements necessary for life, particularly amino acids. More recently, it has been shown that these conditions could also have led to the formation of nucleotides, the basic constituents of modern genetic material (DNA and RNA). Since all forms of life on earth, from the most rudimentary bacteria to evolved animals such as human beings, live and reproduce using the same DNA and RNA codes, the emergence of these molecules can be considered to be the most important stage in the history of life on earth.

But it was the creation of structures able to use the information contained in the structure of DNA as a means of self-replication that really gave impetus to the evolution of life—the point of departure for the living world we know today. At first glance, such a diversification of life may seem unlikely because, from the perspective of a human lifespan, it is hard to imagine the number of events that can take place over hundreds of millions of years. Like the evolution of the brain, as discussed in the preceding chapter, the evolution of life is a very slow process involving the gradual development of efficient and reliable systems to cope with changes in the environment (natural selection). When they prove especially useful, life shows an amazing ability to maintain these systems for long periods of time, as witnessed by the use of DNA as the universal code of life for the past 4 billion years. The broad outlines of the development plan for multicellular organisms, which appeared approximately 500 million years ago, are another example of this "conservation instinct." Even though at the outset this plan produced only primitive invertebrate animals, its broad outlines have been preserved up to the present: for example, you have perhaps wondered why all the insects and animals around us have symmetrical shapes, with half their body being the mirror image of the other half. This symmetry is due to the emergence of the Hox genes 500 million years ago. These genes determine the location of body organs and limbs in relation to each other along the axis that runs from one end of an

< Animal biodiversity (from the top left) Sea urchin; tunicate with brittle star wrapped around it; praying mantis; dragon fish; clownfish; clam; female wolf spider (family *Lycosidae*); king vulture (*Sarcoramphus papa*); black eagle; green python; honeybee; peacock; seahorse; Christmas tree worm (*Spirobranchus giganteus*); Natal forest tree frog (*Leptopolis natalensis*); agama lizard

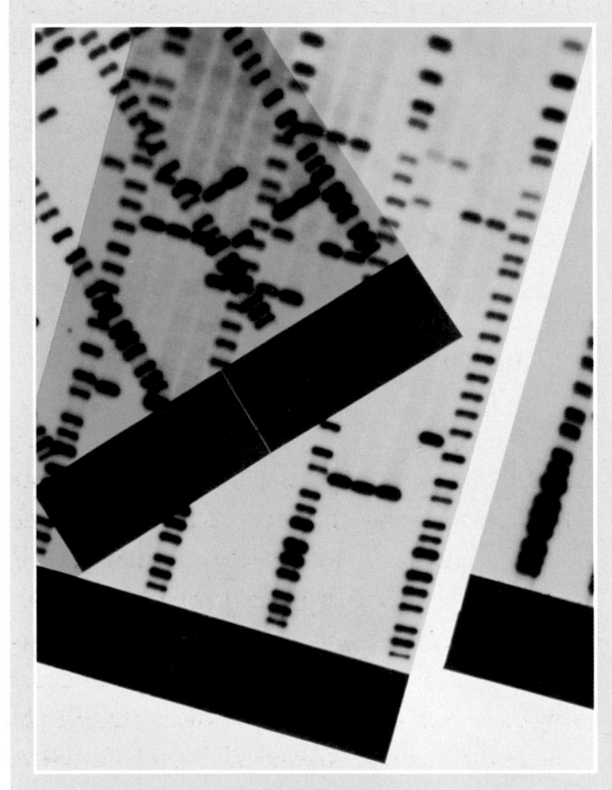

organism to the other (the anteroposterior axis). The survival benefits conferred by these genes were such that they have been meticulously preserved up to the present, hence the symmetrical appearance of all insects and animals on the planet.

The history of life is thus essentially that of the identification of efficient systems able to adapt to the difficulties imposed on them by the environment. We are here because of repeated successful attempts over the course of millions of years to create the biological complexity necessary for the evolution of life. Each step that allows a species to successfully overcome these difficulties is carefully stored in the memory of its DNA, the template of our heredity. To evolve, nature does not reinvent what already works: it fine-tunes these systems and uses them in the most efficient manner possible to maximize the propagation of the species through these evolutionary advantages.

It goes without saying that chance also plays a determining role in the evolutionary process. Nothing, after all, predestined this first primitive cell to cause the emergence of the human species 4 billion years later! Nearly 99% of all species that have seen the light of day in the course of history have vanished from the planet because of the numerous climatic upheavals and other events that have marked the earth's history. Mass extinctions during the Permian (250 million years ago) and Cretaceous (65 million years ago) periods did not necessarily select the most evolved species, but instead those that could best adapt to the enormous changes caused by the relevant catastrophes. Had even one of these species escaped extinction, life on earth may well have looked totally different. For example, if dinosaurs had been able to avoid extinction in the Cretaceous period, the earth today would very likely resemble a kind of "Jurassic Park," but without primates or humans...

^ DNA gel

on earth almost 4 billion years ago (see Figure 1).

This diversification of life did not happen overnight: for more than 3 billion years, or approximately five-sixths of the entire history of life on earth, unicellular organisms were the only life-form on earth. If we transpose the history of the 4 billion years of life on earth onto a scale of one year with 365 days beginning on January 1st, unicellular organisms were the only inhabitants of the planet until the appearance of the first invertebrate "animals" on November 6th. The first plant species would have appeared on November 20th, fish on November 24th, insects on November 29th and the first mammals on December 25th. As for the first humans, they made their appearance only on December 31st, 30 minutes before the end of the year!

Thanks to Darwin's intuition and observations, we now know that the evolution and diversification of life did not occur by chance, but rather because of an implacable law of nature, called natural selection: the organisms most able to adapt to the pressures arising from environmental changes have a better chance of survival and are more likely to produce many offspring. Conversely, a species that is not flexible enough to deal with these pressures will see its population decrease and, in the long term, will disappear. The law of the survival of the fittest is absolutely ruthless: it is estimated that as a result of the many upheavals our planet has experienced since it formed (metorites, volcanic eruptions, continental drift, ice ages, etc.), nearly 99% of all the species that have existed since life first appeared are now extinct. The history of life is also therefore the history of death.

Death, the source of life

Even for the "victorious" species, which have succeeded in overcoming challenges and to which we owe the profusion of life on earth today, death remains an event inextricably linked to life. Even organisms as simple as bacteria and yeasts, which reproduce themselves simply by splitting their single cell to form two daughter cells, are not really immortal: we now know that in the course of this division, one of the two daughter cells ends up with more damaged structures than the other, and this cell eventually jeopardizes its descendants' survival. Sooner or later, everything

> The Tasmanian tiger, now an extinct species.

A question of equilibrium

From the point of view of physics, life is what is known as an open thermodynamic system—a system that constantly exchanges energy with the outside environment. To get an idea of what it costs to maintain such a system, imagine what would happen if you heated your house in very cold weather while keeping some windows open. In such conditions, maintaining a constant temperature would require the heating system to always be on, to compensate for the cold air coming in through the openings. Aside from the fact that this will be very expensive, even the most efficient heating system cannot work forever. Sooner or later there will be a breakdown, the heat source will cut off and the temperatures in the two areas, inside and outside, will eventually balance out and become the same. In the same way, maintaining cellular functions requires a constant supply of energy to remain organized in the face of external disorder, and this continued effort can ultimately only lead to the cell's death. Life is a state of non-equilibrium in relation to the external environment, a state that goes against the natural tendency of matter to reach equilibrium. According to the laws of thermodynamics, death is the return to equilibrium. It is, as a result, inevitable.

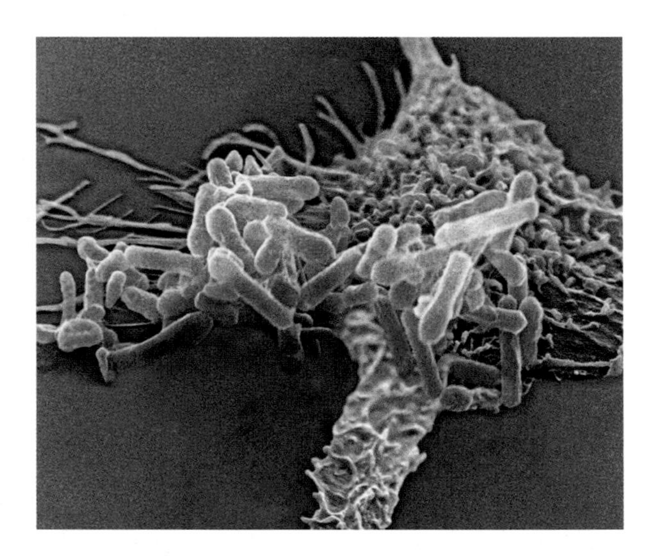

that is alive will die. The only way to continue life's adventure is to manage to ensure the reproduction of the species before dying.

The inseparable relationship between life and death is due to the enormous energy investment required to maintain life. Life consists of a series of biochemical reactions using the energy in the environment to create and maintain complex and orderly structures that are able to replicate themselves by means of cells. Maintaining order in this way is extremely costly because it demands a constant supply of energy in order to counter the fundamental tendency of matter toward a state of disorder. With time, however, this expenditure of energy causes significant damage to cells and cannot successfully maintain order (see box, left).

^ *Pseudomonas* bacteria (red) being attacked by a macrophage (purple)

> Human biodiversity

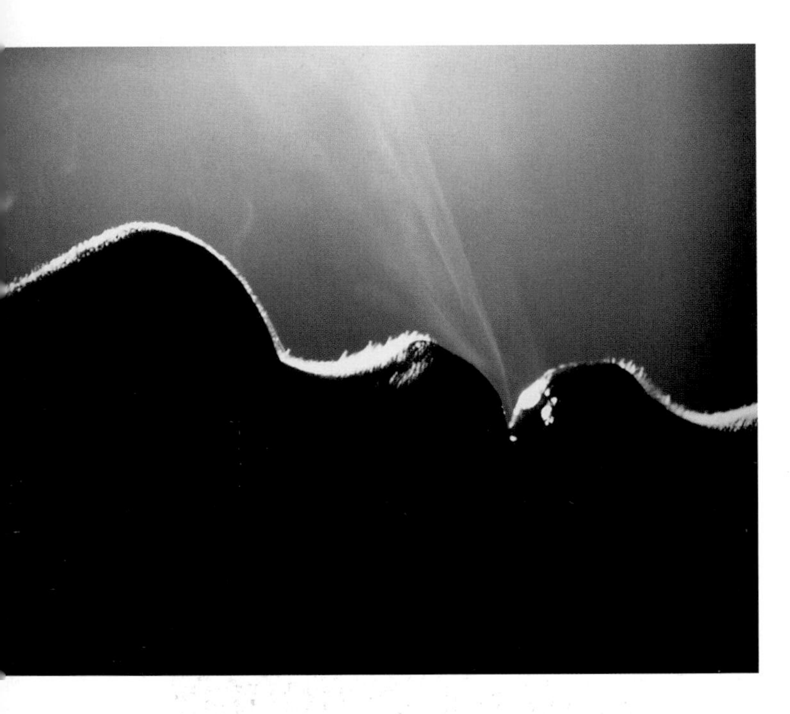

Hence, whether from the point of view of physics, biology or evolution, immortality is not really a viable option! This is precisely why, from the beginning, it was not the ability of cells to survive during long periods that was the impetus behind the evolution of life, but rather their ability to reproduce quickly, before accumulated damage led inevitably to death. Because it produces new and younger organisms better able to adapt to a changing external environment, reproduction is the real driving force behind evolution. If, instead of developing its reproductive potential, the first cell had focussed its energy on battling the ravages caused by the wear and tear of time in order to become immortal, we would have never seen the light of day. Paradoxical though it may seem, it is only because of death that life has been and still is able to reach its full potential.

Taming oxygen

Given the amount of energy that must continually be expended to maintain life, it is not surprising that the emergence of more evolved lifeforms, consisting of many cells, was only made possible by the creation of high-efficiency metabolic mechanisms able to provide large quantities of this precious energy. Very early in evolution, adenosine triphosphate, better known by the abbreviation ATP, became the living world's universal fuel. In the case of the first bacteria, which appeared in an era when the atmosphere was essentially devoid of oxygen, ATP had to be produced by fermentation. Although this process is adequate to support cellular functions (several microorganisms have maintained this means of energy production up to the present), it nonetheless is clearly insufficient to ensure the survival of a complex organism composed of several billion cells.

Oxygen was to be the catalyst for the explosion of life on earth, and the emergence of more evolved life forms coincided with a dramatic increase in oxygen molecules in the earth's atmosphere. The earliest traces of oxygen appeared around 2.5 billion years ago, as a result of the metabolism of cyanobacteria (blue-green algae) that released it as "waste" while producing molecules essential to their survival through photosynthesis. This photosynthesis activity, combined with that of plants after they had colonized the earth's surface, led to a slow but dramatic increase in the level of atmospheric oxygen in the millions of years that followed (see Figure 2).

This increase in oxygen went hand in hand with what was really an explosion of life on earth, as shown by the emergence of several types of invertebrate animals, notably the Ediacaran fauna (the oldest fossils of complex organisms, dating back 565 million years and discovered in the Australian hills of the same name). This rapid development was directly linked to an enormous improvement in oxygen-based energy production. For example, whereas the fermentation of a molecule of glucose in simple unicellular organisms produces only two molecules of ATP, the metab-

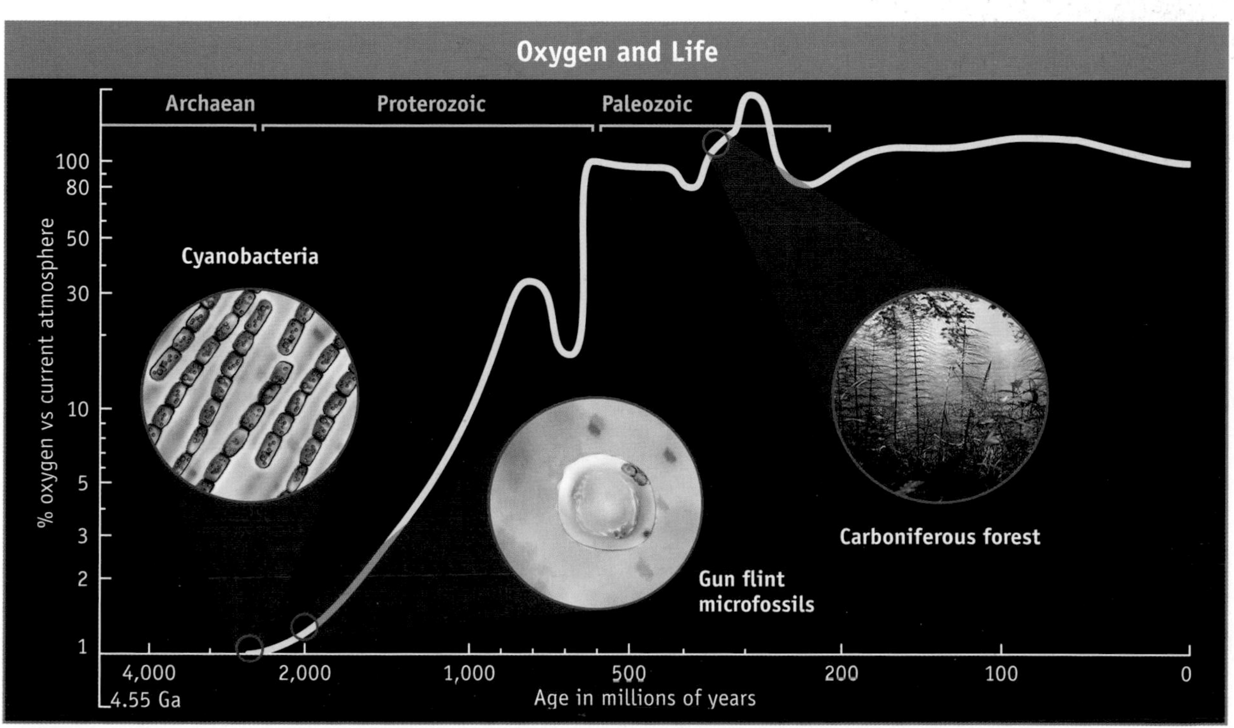

Figure 2

olism of this same molecule of glucose in the presence of oxygen produces 36 ATP, or 18 times the output!

This increase in efficiency is a direct consequence of the happiest marriage in the entire history of life on earth, an "old couple" that got together more than 2 billion years ago, the union of a bacterium able to transform oxygen into ATP and a primitive cell that was, on its own, unable to use the new gas now found in the environment. Life as we know it today could have never existed without the union of these two primitive life-forms: by enabling cells to transform oxygen into ATP efficiently, these bacteria supplied the fundamentals needed for the evolution of more complex life-forms that required an increased energy supply in order to live and reproduce.

In "modern" cells, called eukaryotes, these ancient bacteria take the form of mitochondria, found in both animals and plants, and chloroplasts, found only in plants.

Mitochondria still have their own DNA, able to code for certain proteins and RNA (in human beings, no fewer than 37 mitochondrial genes participate in cellular functions). Unlike nuclear DNA, which is inherited from both parents, mitochondrial DNA is only passed on by the mother, and it is possible to use this characteristic to trace the origin of our species. Based on currently avail-

^ Trilobite fossil dating from about 500 million years ago.

able data, it is thought that all human mitochondria may have a common African origin, a "mitochondrial Eve" who lived in Ethiopia, Kenya or Tanzania roughly 150,000 years ago.

The main role of mitochondria is to act as power stations for the production of ATP. In plant chloroplasts, it is the sun's electromagnetic energy that is converted into chemical energy. This transformation of one form of energy into another marks the origin of life on our planet (see Figure 3). In mitochondria, it is the chemical energy of proteins, sugars and lipids that is converted into fuel—ATP. The mechanisms at work in producing this energy are so complex that for generations the greatest biochemists tried to understand how they worked and could not make head nor tail of them. It is only owing to the genius of a few, notably British scientist Peter Mitchell (1978 Nobel Prize-winner), that we finally understand the big picture (see Figure 4).

This model, called the chemiosmotic model, explains how the biochemical energy in the energy-rich molecules found in food (sugar, lipids, proteins) or solar energy, captured by vegetable pigments in plants, is used to generate a current of electrons in the mitochondrial membrane that, in turn, creates an electrochemical gradient of protons on both sides of the mitochondrial membrane. This gradient is then used by an enzyme (F_0F_1ATPase) to synthesize precious ATP.

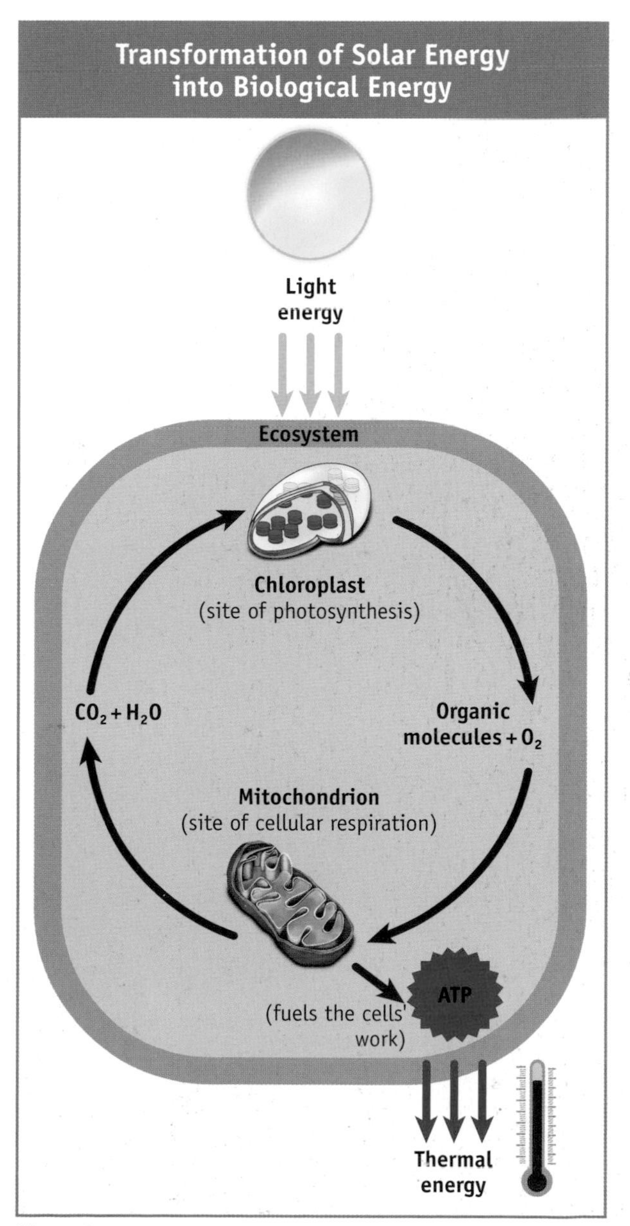

Figure 3

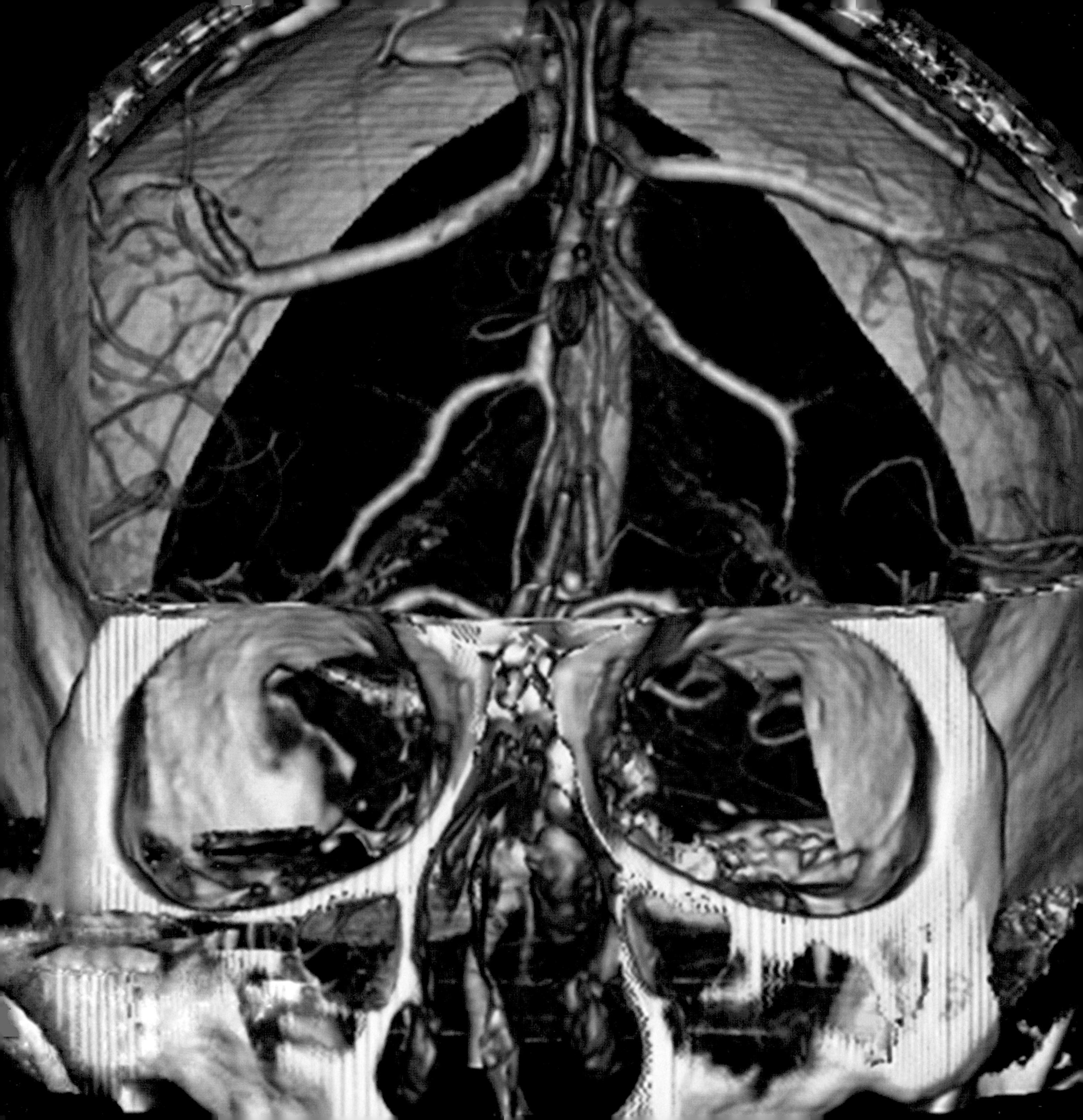

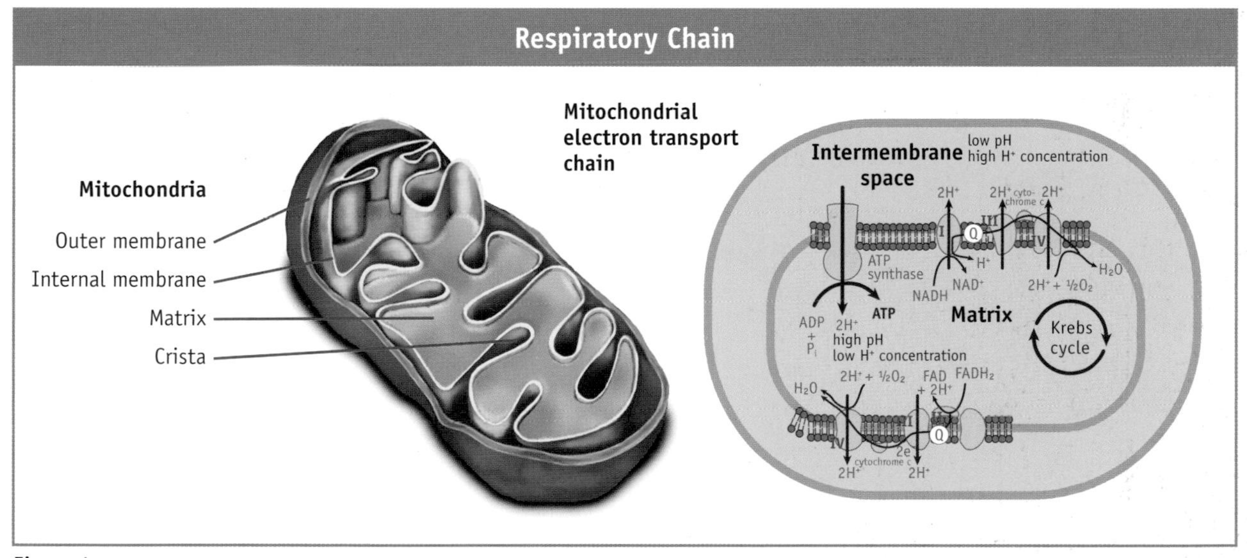

Figure 4

This entire process is called cellular respiration and leads to the release of CO_2 and to oxygen consumption. $C_6H_{12}O_6 + 6\ O_2 \rightarrow 6\ CO_2 + 6\ H_2O$ + energy (ATP and heat).

Oxygen is absolutely essential to the lives of animal species simply because cells have universally adopted this method of energy production in order to function. Cellular respiration, occurring in all cells in the human body, means that a system had to be put in place to deliver oxygen to all cells, even those hidden deep inside our tissues and not exposed to atmospheric oxygen—hence the blood system, the vehicle for red blood cells. Red blood cells contain hemoglobin, a pigment whose principal characteristic is its very high affinity for oxygen. Carried by the blood through a network of thousands of miles of capillaries, precious oxygen is thus delivered to all the body's cells. Respiration is usually viewed as a physical and macroscopic phenomenon, in which the movement of the diaphragm enables the lungs to breath in air containing roughly 20% oxygen, but this macrophenomenon is just the evolutionary result of true metabolic respiration, which is mitochondrial in origin.

It is often said that a system is only as strong as its weakest component. In the case of cells, this heavy dependence on oxygen means that

< Cerebral angiogram showing the blood vessels in the brain.

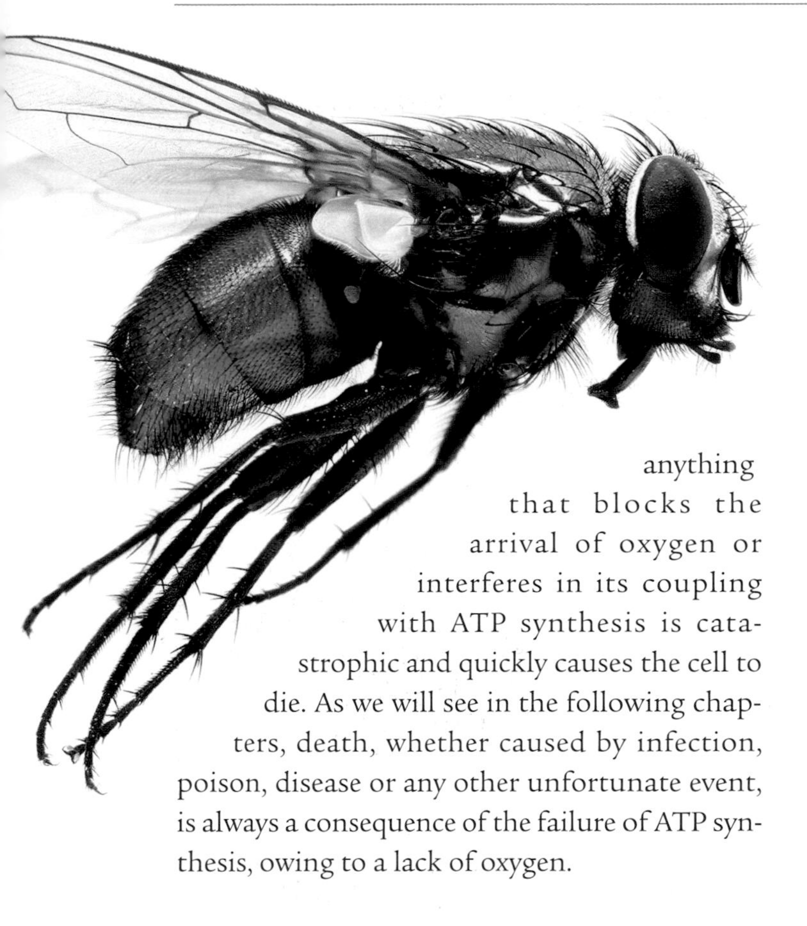

anything that blocks the arrival of oxygen or interferes in its coupling with ATP synthesis is catastrophic and quickly causes the cell to die. As we will see in the following chapters, death, whether caused by infection, poison, disease or any other unfortunate event, is always a consequence of the failure of ATP synthesis, owing to a lack of oxygen.

Shaping life

In spite of oxygen's ability to sustain the function of more evolved organisms, life would have never reached the degree of complexity we know today without the active participation of death.

We may not always be aware of it, but all animals, even those we often consider "inferior," like insects, fish or reptiles, are really marvels of evolution, the functional organization of several million cells that make it possible for these animals to feed themselves, move around and clearly perceive their surroundings. Such complexity would of course be impossible if all cells were identical; it is only because of their specialization—that is, the acquisition of certain properties that enable them to accomplish a specific task for the benefit of the whole organism—that each of these animals acquires its shape and specific way of life.

This specialization, known as cellular differentiation, begins in the very first stages of embryonic development. The vast majority of multicellular animals (with the exception of certain species like sponges and corals) are triploblastic, meaning that once the egg has been fertilized by the sperm, the embryo divides into three distinct layers (ectoderm, mesoderm and endoderm) that then turn into the array of specialized cells found in the animal. For example, the outer layer of the embryo (the ectoderm) is responsible for the formation of the nervous system's neurons as well as skin cells; the intermediate layer (the mesoderm) is involved in the development of muscles, kidneys, reproductive organs, etc.; and from the internal layer (the endoderm), the digestive system as well as several other types of cells (lungs, thyroid and pancreas) are formed (see Figure 5). That a single fertilized egg can

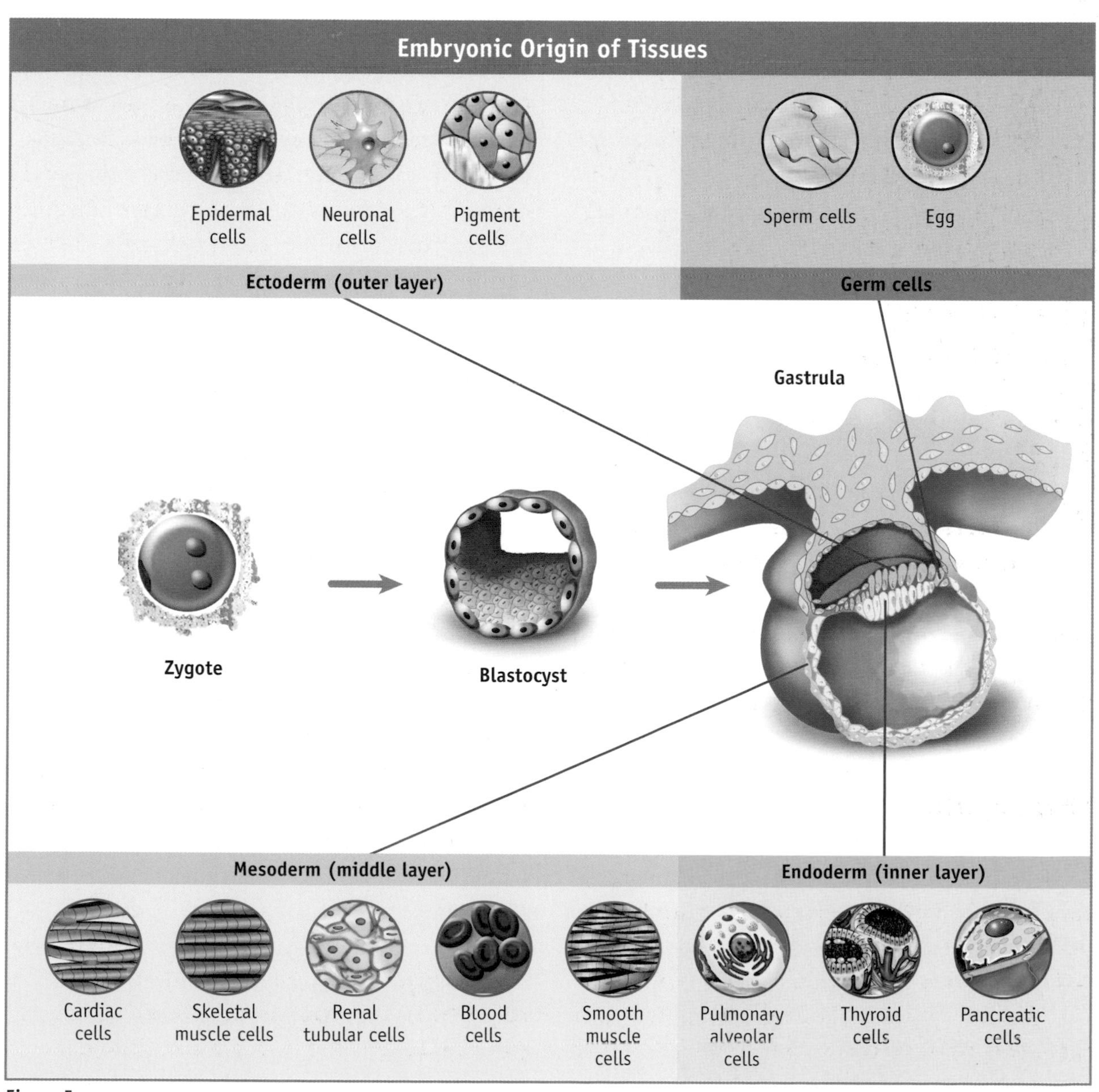
Embryonic Origin of Tissues
Epidermal cells
Neuronal cells
Pigment cells
Sperm cells
Egg
Ectoderm (outer layer)
Germ cells
Gastrula
Zygote
Blastocyst
Mesoderm (middle layer)
Endoderm (inner layer)
Cardiac cells
Skeletal muscle cells
Renal tubular cells
Blood cells
Smooth muscle cells
Pulmonary alveolar cells
Thyroid cells
Pancreatic cells

Figure 5

give birth to such a variety of distinct cells, able to carry out functions as specialized as the transmission of nerve impulses, the perception of light and the digestion of food, is without doubt one of nature's great masterpieces.

This general organization took shape around 575 million years ago and has since been the framework of natural selection, within which multicellular species have evolved in response to changes in the environment. Although it often appears dramatic, evolution rarely requires the creation of a new structure, but instead demands a patient reorganization of available elements to cope in the most effective way with the challenges posed by the environment. For example, while a human being's forearm, a bat's wing, a seal's pectoral fin or a horse's hoof may seem to be completely different, all of these limbs are in fact

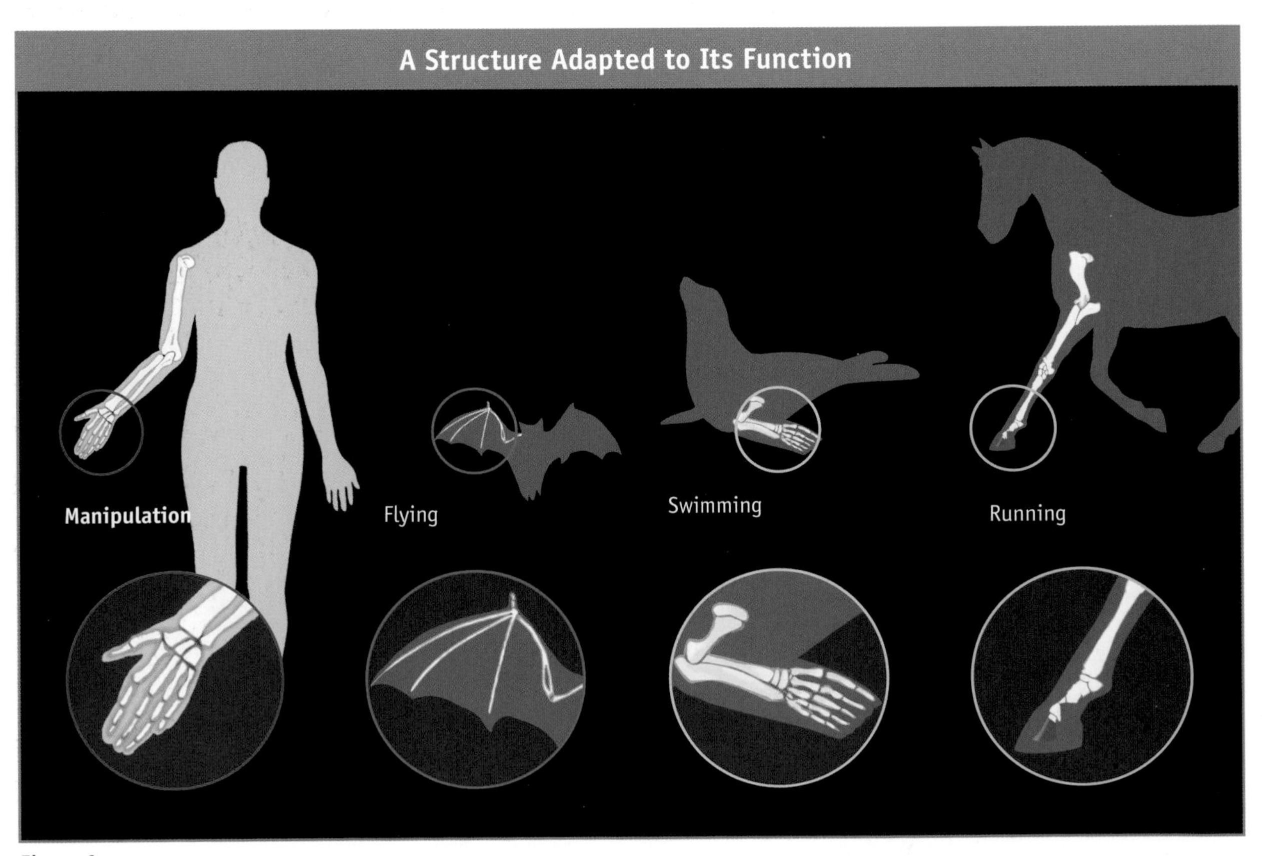

Figure 6

homologous structures, made up of bones inherited from a common ancestor that are simply arranged differently to perform particular physiological functions (see Figure 6).

The processes that make such complexity possible are far outside the scope of this book, but, even without any knowledge of biology or the sciences in general, it is still possible to understand intuitively the extent to which an animal as evolved as a human being shares a common heritage with other "inferior" animals. For example, simple observation of the morphology of the embryos of several species during the first stages of their development shows how similar they are (see Figure 7). Even with species as different as mice and human beings, it is virtually impossible to tell which embryo is which in the early stages of development!

Programmed cell death

Cell diversification and specialization in more evolved organisms may in some respects be compared with that of a modern society, where labor specialization is associated with more complex social organization. Despite the numerous advantages this form of civilization offers, a well-

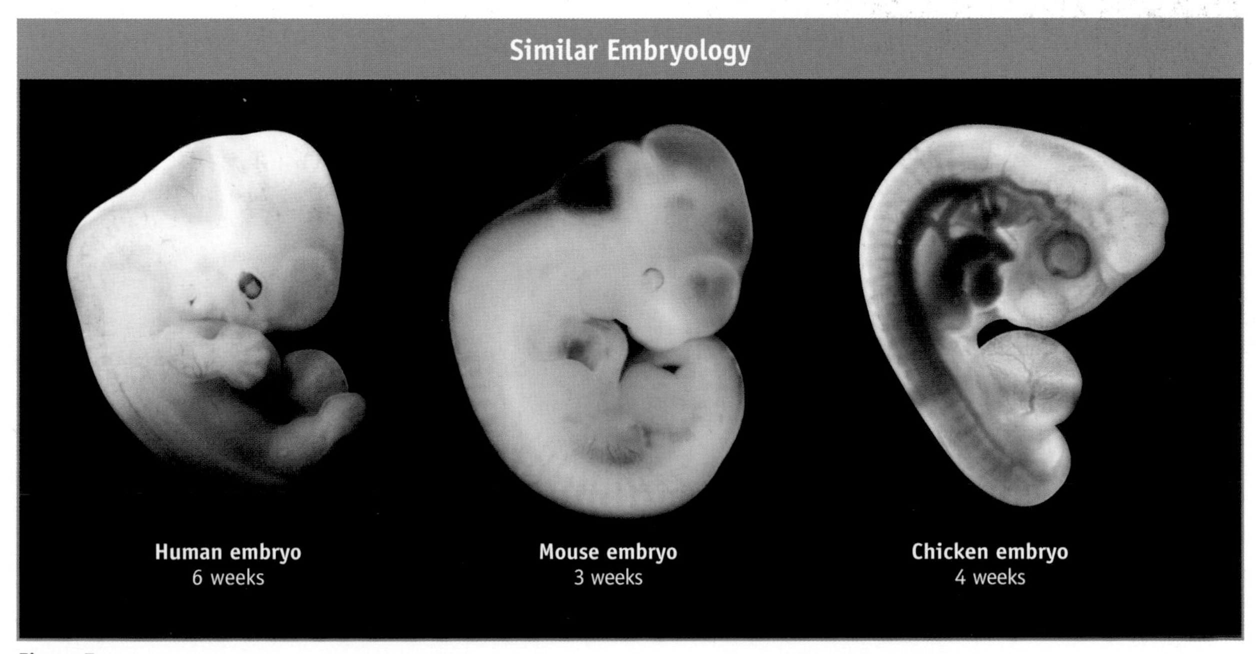

Figure 7

ordered structure cannot be maintained without conflict, and certain strict rules must usually be imposed to preserve order.

In the case of development, specialized structures could not be formed without eliminating superfluous cells, whose presence is incompatible with the proper functioning of the organism. This eradication is made possible by the presence inside each cell of a very sophisticated self-destruction mechanism that is able to set in motion a kind of a "ritual sacrifice" whenever needed. Known as apoptosis, this cell sacrifice, or programmed cell death, involves a complete dismantling of the cell by death enzymes, the caspases, which are molecular "scalpels" that methodically shred the components of the cell. In response to an "execution order" issued by the surrounding cells or upon detection of irreparable damage that could interfere with the normal functioning of the cell, an amazing series of events intended to eliminate the cell is set in motion (see Figure 8). For example, if the mitochondria detect any change in a cell's normal functioning, they release into the cell one of the proteins normally involved in the synthesis of ATP (cytochrome c). The abnormal presence of cytochrome c is immediately taken as a trigger for cell sacrifice, an alarm signal that begins to activate a series of caspases and launches a countdown to the execution of the cell, easily observable in the formation of blisters on its surface. Mitochondria are sources of life, but they are also central to cell death.

Apoptosis plays an essential role in shaping the body's organs during development: for example, as brain structures are being formed in the embryo, neurons that have not come into contact with other neurons to participate in the transmission of nerve impulses are eliminated. In the same way, the individualization of toes and fingers in human beings is due to the highly localized activation of the apoptosis process that causes the destruction of interdigital cells.

Dying little by little

The process of apoptosis is extremely important in all living species. Every day, in absolute anonymity, close to 10 billion of our cells, having become inefficient, die in the apoptosis process, with each one of these cells, luckily, being replaced right away by a new functioning cell. The rate of death and rebirth varies considerably from one cell to another. Whereas the lifespan of a cell lining the intestinal wall is no more than 5 days, the number of cells in our nervous system does not really vary much throughout our life (see Figure 9). This constant renewal means that the majority of the cells in our body are less than 10

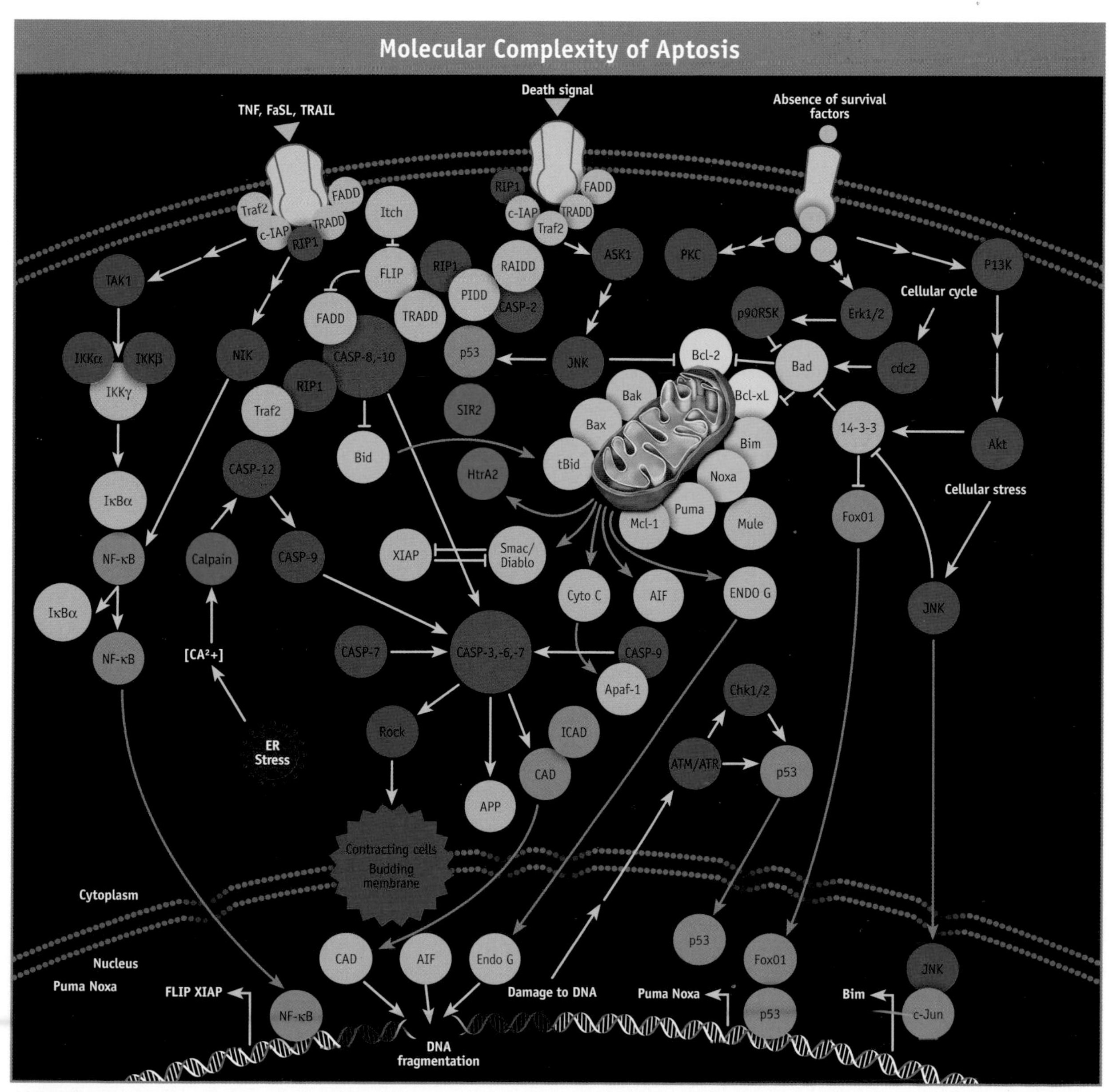
Molecular Complexity of Aptosis
TNF, FaSL, TRAIL
Death signal
Absence of survival factors
Traf2
c-IAP
RIP1
FADD
TRADD
Itch
FLIP
RIP1
RAIDD
PIDD
CASP-2
ASK1
PKC
P13K
TAK1
Cellular cycle
p90RSK
Erk1/2
FADD
TRADD
IKKα
IKKβ
IKKγ
NIK
CASP-8,-10
RIP1
p53
JNK
Bcl-2
Bad
cdc2
Traf2
SIR2
Bak
Bcl-xL
Bax
14-3-3
Akt
Bim
Bid
HtrA2
tBid
Noxa
IκBα
CASP-12
Puma
Mcl-1
Mule
Fox01
Cellular stress
NF-κB
Calpain
CASP-9
XIAP
Smac/ Diablo
Cyto C
AIF
ENDO G
JNK
IκBα
NF-κB
[CA²+]
CASP-7
CASP-3,-6,-7
CASP-9
Apaf-1
Chk1/2
Rock
ICAD
ER Stress
ATM/ATR
p53
CAD
APP
Contracting cells
Budding membrane
Cytoplasm
p53
Nucleus
CAD
AIF
Endo G
Fox01
JNK
Puma Noxa
FLIP XIAP
NF-κB
Damage to DNA
Puma Noxa
p53
Bim
c-Jun
DNA fragmentation

Figure 8

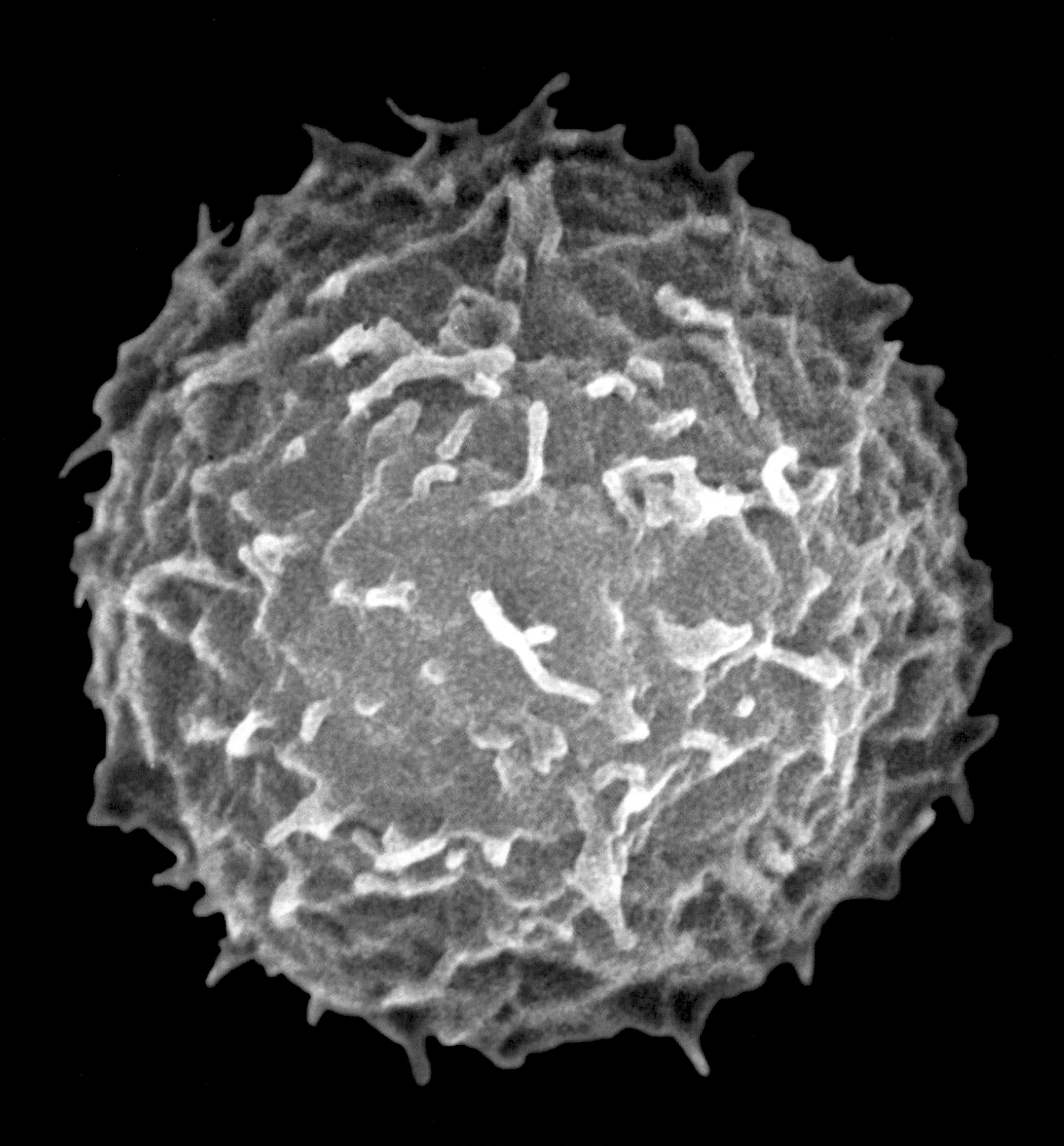

Normal white blood cell

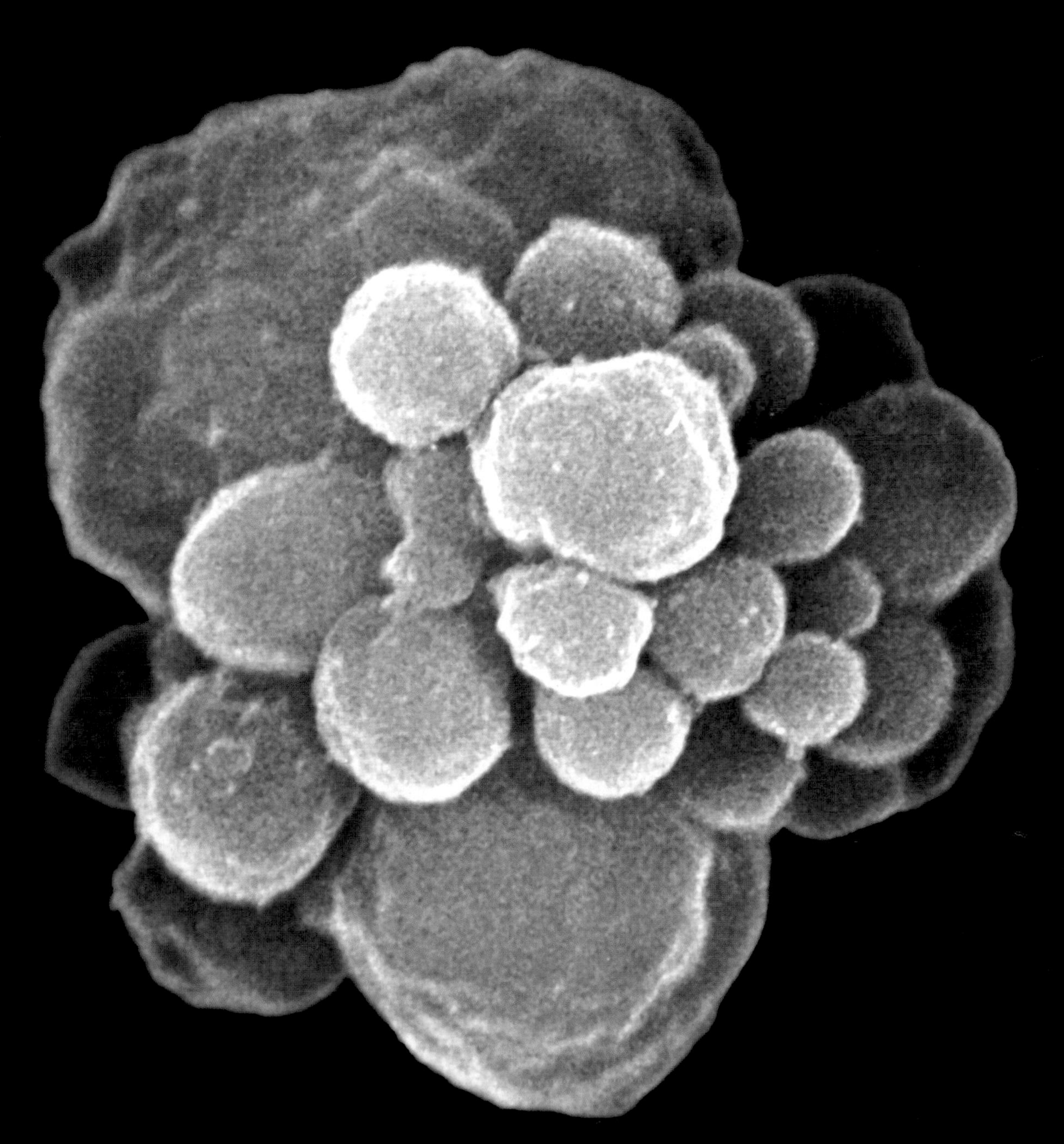

White blood cell undergoing apoptosis

years old and are therefore much younger than we are. So it's normal to feel younger than your age!

In spite of its efficiency, this renewal potential is limited and decreases over time, resulting in physiological functions gradually deteriorating.

Cells with a Short Lifespan

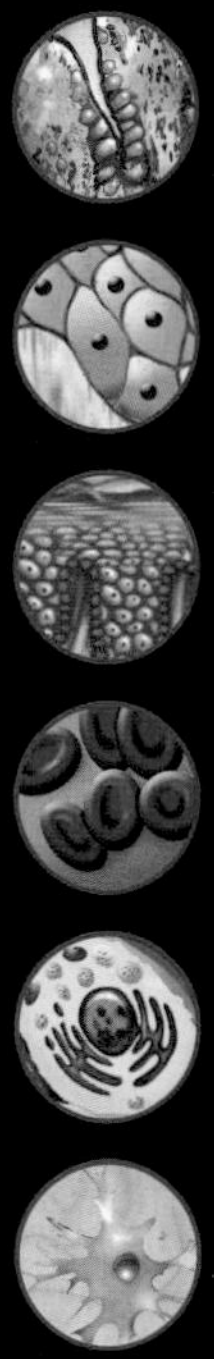

Cell type	Average lifespan
Epithelial (inner intestinal wall)	5 days
Retina	10 days
Skin	21–28 days
Red blood cells	120 days
Pneumocytes (lung)	400–500 days
Neurons	> 60 years

Figure 9

This long series of "little deaths" that occur throughout life eventually reaches a point of no return, the inevitable result being an increasingly significant loss of vital functions and, ultimately, the death of the organism. When all is said and done, the fact that we eventually die is due to the fact that we are dying a little every day.

Obviously death is not a fate exclusive to the human species; all living beings, be they plants, insects, fish, birds or more complex animals, are born, grow up and die at their own rhythm. Biologically speaking, the molecular and cellular phenomena that end the life of every living organism are exactly the same as those at work in human death. Our death is neither an anomaly nor a fate unjustly imposed on the human species, but rather the only logical conclusion to life. However, we perceive the passage of time and the inevitability of death, and we use our intellectual faculties to reflect on the meaning of life and death. This strength, which has allowed us to become the dominant species on earth, may also be our Achilles heel, a source of anxiety that can poison our existence. Thinking is what causes a fear of death.

> Louis Édouard Fournier, *The Funeral of Shelley* (detail)

Chapter 3

Living with the Awareness of Death: Between Hope and Fear

> I don't want to achieve immortality through my work.
> I want to achieve it through not dying.
>
> Woody Allen (1935–)

All living species try to avoid death. A simple bacterium placed in contact with a toxic substance uses a complex mechanism to change its path to avoid the danger; a plant attacked by a parasite triggers the synthesis of toxic substances to neutralize the threat; an antelope standing beside a pond is constantly on the alert for a potential predator crouching in the high grass, torn between the desire to drink and the fear of dying. Living beings are always "anxious" about death, an instinct that is inextricably bound up with the very presence of life on earth.

In animals, the detection of a danger (the smell of smoke, the sight of a predator, a shotgun noise) by the sensory organs (nose, eyes, ears) triggers a maximum alert in the brain, which initiates a series of extremely complex processes called the fight-or-flight response. By stimulating the adrenal glands, the brain causes action hormones like adrenalin to be released into the bloodstream. This increases the respiratory rate, heartbeat and supply of oxygen to the tissues, as well as raising the brain's level of alertness and attention. These changes all make it possible to either fight or quickly flee from the danger. This is biological stress, essential for the survival of an individual and hence that of the species. This anxiety in the face of potentially deadly dangers is not conscious, but it serves a reproductive function, enabling the animal to live long enough to pass on its genes to its descendents and protect its offspring while they are still vulnerable.

Where does the actual awareness of death come from? In the 1970s, psychologist Gordon Gallup developed a very simple test to measure an animal's self-awareness. It involves putting a colored but odorless mark on the animal's head and observing whether the animal is able to react to the presence of the mark—in other words, to recognize its own image in a mirror as an image of itself. Based on this criterion, very few animals belong to what could be called the "elite" of the animal kingdom: the apes, such as chimpanzees, bonobos and orangutans, pass the test without difficulty, as do dolphins, killer whales and some elephants. Even some birds in the Corvidae family, such as the magpie, can recognize themselves, which goes to show that being "bird-brained" might not be as bad as some would have us believe!

As well as indicating an ability to reason and an intelligence that are clearly above average, this self-awareness in animals also goes hand in hand with the appearance of the first signs of awareness of the reality of death. Indeed, a number of biologists have observed behaviors in many of these "conscious" animals that express definite distress in the presence of the dead bodies of members of their species, notably the holding of "funeral vigils," during which dolphins, chimpanzees and elephants remain near their dead companions for several days. The awareness of death thus seems to be an inevitable consequence of the development of intelligence, the expression of a sensitivity made possible by a more acute perception of the world around them and events taking place in it.

Confronting death

The phenomenal development of the human brain over the course of evolution has coincided with the emergence of an awareness of the precariousness of life found nowhere else in the living world. Indeed, this heightened awareness of death is often considered to be our species' "signature," the main characteristic distinguishing us from other animals. As has been observed in the more evolved mammals, awareness of death is first and foremost a cultural perception; in other words, it is the death of others that brings us face to face with our own mortality. As Russian novelist Ivan Bounine (1870–1953) noted, if we had been born and spent our entire lives on a desert island, we would not even have suspected that death existed.

In its simplest form, the passage from life to death was basically viewed as a loss of energy that could be thwarted by "absorbing" the energy of the deceased. In fact, a funeral practice that appeared very early in human history consisted of eating the dead person's body. This endocannibalism was revealed by signs of dismemberment and chewing on human bones dating from prehistory (100,000 years ago or more). Horrible though it may be, ritual cannibalism has been observed again and again in several parts of the world. It is no doubt the expression of a basic instinct, for such behaviors, although rare, have also been observed in other primates. This kind of cannibalism, while obviously uncommon today, has nonetheless been regularly practiced, even in recent times (see box on pp. 73–74), not to mention that it has sometimes been the only means of survival for people stranded in extreme conditions. For example, following the shipwreck of the French ship *La Méduse* off Mauritania in 1815,

^ Skull of a Neanderthal

High-Risk Cannibals

The Fores tribe, which still lives today on the high plateaus of Papua New Guinea, practiced a ritual funerary cannibalism during which the men ate the deceased's muscles, symbols of strength, while the internal organs and the brain were steamed and reserved for the women and children. Unfortunately, a significant proportion of those who ate the organs and brain became ill with kuru, a disease whose symptoms include trembling (kuru means "trembling with fear" in their language) and a short-lived euphoria followed by a dramatic deterioration in a number of neuromotor functions (incontinence, difficulty swallowing, imbalance). In the 1950s, autopsies revealed that these diseased individuals showed considerable brain damage, with the brain looking positively sponge-like (hence the term spongiform encephalopathy). Without knowing what infectious agent was responsible for this disorder, researchers suspected it was linked to the cannibal rituals, and, indeed, when these practices were stopped, there was a significant decline in cases. Today we know that this disease is similar to sheep scrapie and mad cow disease (bovine spongiform encephalopathy), two diseases caused by the transmission of infectious agents called prions. In the latter case, a form of animal cannibalism is at the root of the disease, which originated in the livestock industry and has been linked to the presence of cattle byproducts in cattle feed.

While they have fortunately become extremely rare, cases of cannibalism do occasionally come to light and never fail to arouse a mixture of astonishment and disgust. Among the most famous are Issei Sagawa, the "Japanese cannibal," who was arrested after having devoured 15 pounds (7 kg) of the flesh of a young Dutch girl in order to "absorb her energy"; Anna Zimmerman, who

< Théodore Géricault, *The Raft of the Medusa* (detail)

> "Dismembering and Cooking an Enemy," from *Americae Tertia Pars* by Théodore de Bry.

murdered her boyfriend, dismembered him and froze the pieces, which she then ate over a period of time with her children; and, lastly, Armin Meiwes, the "Cannibal of Rotenburg," who found his victim by simply putting an advertisement in a newspaper: "Seeking a young well-built male for eating. Interest in cannibalism and killing required." After having cut off and eaten (with his consent) the penis of the "chosen one" (Bernd Jürgen Brandes), Meiwes killed him and sliced off 65 pounds (30 kg) of flesh that he then froze so he could eat it at regular intervals. "With every mouthful, my memory of him grew stronger," Meiwes stated. During his trial, Meiwes claimed he felt more stable since he had eaten his victim, quite apart from the fact that he spoke much better English, which he attributed to the fact that Brandes spoke it. In 2006 Meiwes was sentenced to life imprisonment; his cellmates can sleep without fear, however, because he is believed to have become a vegetarian at the end of 2007.

10 or so soldiers and sailors were only able to survive by eating their dead companions.

However, this very primitive view of death rapidly became more complex (and luckily so!). In fact, very early in human history, most cultures developed more or less complex symbolic rituals to mark in a solemn way the departure of the dead from the community of the living. It is obviously impossible to know with precision the exact nature of these early rituals, but the intentional burial of the deceased was no doubt a way of marking the passage from life to death. Archaeological discoveries date such graves from the dawn of humanity. Over time, respect for the dead increased, as shown in the construction of ever more majestic funerary monuments (dolmens, tumuli, cairns) that can still be admired in many parts of the world. As in nature, where it is a universal symbol of birth and life, the earth is also a powerful symbol of death; the return of the body to the earth marks the end of a cycle and the promise of a new life.

One of the most fascinating aspects of many prehistoric graves is the presence of symbols suggesting that the return of the deceased to the earth might in fact be the prelude to a

< Image of a horse on the wall in a cave at Lascaux, France.
< A dolmen

new life. Common objects, weapons, food or the skeletons of domestic animals have often been found near buried remains, suggesting a belief in life in another world after death, or at least the hope that this would be so. The nature of this world may vary considerably depending on the era and the culture, but this nevertheless indicates a major change in the perception of death. It was no longer strictly considered to be the end of existence, but rather the point of departure for another life.

The difficulty human beings have in accepting death and their innate tendency to transcend this inevitable outcome by looking for symbols that justify the relevance of life while explaining the apparent absurdity of death is very well illustrated in the Mesopotamian epic of Gilgamesh, one of the oldest documents in human history. Written in cuneiform script on clay tablets over 6,000 years ago, this story tells how Gilgamesh, King of Uruk, is inconsolable following the death of his best friend. He cannot accept the finiteness of existence and leaves in search of the secret to life without end. That one of the first oral traditions to be written down deals with the difficulty of accepting death and the search for immortality shows that the fear of death is a basic human characteristic that has always greatly influenced the way we think—and still does.

Homo religiosus

Nothing better illustrates this constant search for meaning in life and death than the omnipresence of religions in human history. Homo sapiens really is Homo religiosus, for, as far back as we can go into the past, human beings have always invoked the participation of superior powers to give some kind of meaning to both death and the worldly events around them. The origins of this religious temperament may be even older than civilizations themselves,

< The Uruk archaeological site
< Table fragment inscribed with the "Song of Gilgamesh."
< Statues of soldiers at the site of the tomb of the first emperor of China.

since the oldest temple ever uncovered (Göbekli Tepe, in Turkey) was built roughly 12,000 years ago, or several thousand years before the appearance of the first cities. The factors underlying the emergence of religious beliefs remain obscure, but it is certain that the extraordinary amount of work required to build a place of worship like this (it is estimated that it took 200 to 300 years to move stone blocks weighing several tons over several hundred feet) shows that religious experience has held a fundamental place in human culture for a very long time.

As with all other aspects of human life related to culture (art, tool-making, language, cooking), religious practices have varied considerably across different regions of the world and different eras. At the beginning, religions were basically a way of suggesting a rational explanation for totally inexplicable events, especially those related to suffering and grief. Claiming that divine powers were responsible for thunder, earthquakes, droughts or even premature deaths gave meaning to these hardships, as well as hinting that it might be possible to moderate their effects by speaking directly to these powers, whether through incantations, prayers or sacrifices. Establishing a complex religious ritual must therefore entail finding a way to communicate with the gods. Even in very ancient religions like animism (a belief according to which everything that exists, whether it is the tree, the wind, a stone or an animal, has a "soul"), a shaman's ability to act as an intermediary between the earthly world and that of souls and to intercede with them to improve the human condition (success in hunting, healing) is an essential characteristic of this kind of belief.

The development of civilizations played a key role in the emergence of new forms of religion, both around the Mediterranean and in Asia. These religions were initially polytheist, consisting of pantheons of human-like gods with various attributes (lightning, bow, sword) and ruled by a higher god, somewhat in the image of the hierarchical society that was becoming established at the time, with the founding of city-states. These religions became the first form of organized religious experience, particularly with

< The journey of the deceased to the afterlife, from the Egyptian *Book of the Dead*.
< Funeral mask found in Royal Tomb IV in Mycenae, Greece.

the appearance of places of worship (Greek oracles and temples) and priests. However, it was only in about the eighth century BCE that modern religions really began to emerge, first with the appearance of Judaism along the Mediterranean and Vedism in India and then, a few centuries later, with Christianity, Buddhism and Islam. Despite major differences in their rites as well as in the identity of the gods they worship, all of these religions put forth that the ultimate goal of earthly existence is to achieve salvation in the afterlife. In this sense, the emergence of these religions of salvation appears to be a crucial stage in human beings' perception of death. The promise that human life does not end in dust and nothingness but continues in a better world is a powerful message that may lessen fears about the end of existence.

In the monotheist and Abrahamic religions (Judaism, Islam and Christianity), the soul manages to triumph over death and enter eternal life.

> He will swallow up death in victory, and the Lord God will wipe away tears from all faces (Isaiah 25:8).
>
> And the dead shall be raised (Corinthians 15:52).
>
> **The Bible**

> Then, later on, you die. Then, on the Day of resurrection, you will be resurrected.
>
> **Koran, Sura 23:15–16**

However, in most parts of the Far East, the existence of a god does not have the same dominance. Instead, the spirit lives in the person, who must try to achieve purification and the perfection of a life that will in the end bring bliss, nirvana or satori, a state in which desire, tension and anxiety no longer exist and where a person is finally free of earthly concerns. This purification requires a series of cycles, passages or reincarnations that will ultimately free the spirit to leave behind the material world and reach nirvana.

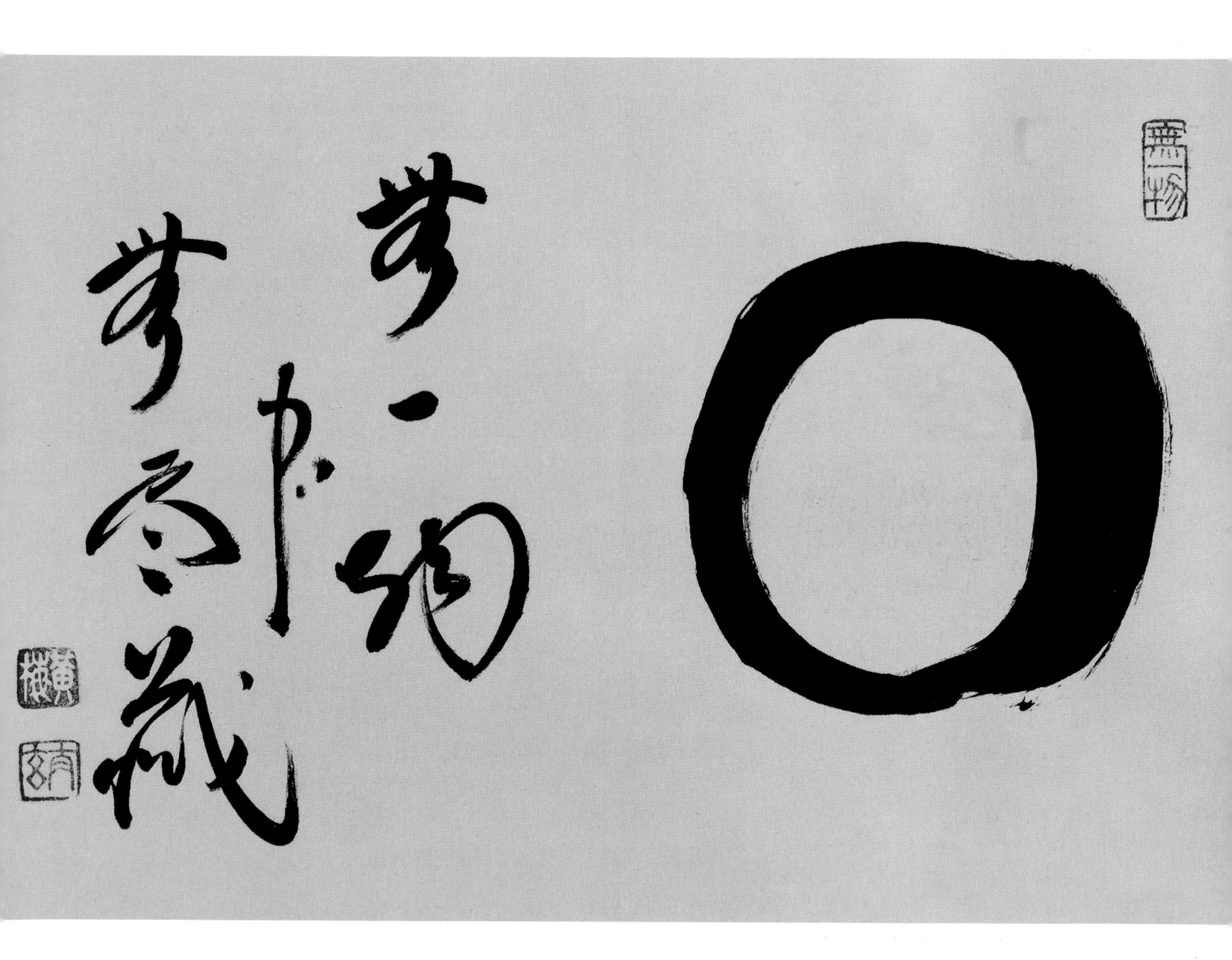

^ The circle *(Enso)* is one of the most important symbols in Zen Buddhism. It represents strength, enlightenment, the universe and emptiness. Because it is so difficult to draw with a brush on highly absorbent rice paper, it symbolizes the concentration of the mind on the present moment because its technical execution is extremely complex. This "expression of the moment" is characteristic of Zen and the importance of concentration. It is one of the many images of the duality between life and death and the fullness of existence, which includes an awareness of death.

The prospect of a life after death has also played a key role in the observance of ethical codes imposed by religions; the fear of being punished and not being able to attain eternal life was a powerful incentive to obey these rules. It is therefore not surprising that the concept of hell should hold a vital place in nearly every religion: whether it is Dante's descent through the nine circles of hell in *The Divine Comedy,* the Muslim Gehenna or the 18 Buddhist hells, the purpose of all of these infernal worlds is to cause terrible suffering (usually by fire) for the souls of those who have committed serious crimes and sins during their earthly lives. The terrifying prospect of burning in hell helped maintain social order by preventing certain deviant behaviors in society. The historical importance of religions suggests that fear of death and ignorance of the physiological processes associated with it have played a crucial role in their emergence and in our perception of death in general.

Dying of fright

Of course, the existence of a postmortem life invisible to the living leaves a great deal of latitude in the interpretation of the events that occur after death. In some cultures, the dead have the right to an independent life, somewhat as if a twin separated itself from the person at

^ Hieronymus Bosch, *The Haywain* (detail)

the moment of death and began another life in a world just for the dead. In others, this life takes the form of ghosts or spirits that coexist with the world of the living and can thus come back and meddle at any time in what the living are doing. This latter view is obviously not at all reassuring. The result is that these beliefs have given rise to a wide range of myths, legends and tales in which ghosts, especially if their death was in any way sordid, can destroy the lives of the living.

The survival of the dead and their ongoing presence in a world parallel to our own has added a fear of the dead to the existing fear of death in general. This fear of the dead can be seen very early in history; for example, among certain American indigenous peoples as well as in several parts of Europe (notably Alsace), the feet of the dead were tied together with string to keep them from coming back to haunt the living. Similarly, during the Bronze Age, graves were often covered with huge mounds of stones (tumuli) around which were dug ditches, perhaps to keep the dead from returning to the world of the living. Precautions designed to keep a dead person from coming back were sometimes taken very soon after death: carrying the deceased out feet first to keep him from "looking" into the house and closing the windows after a death to block his return, to name just two. We may smile at these beliefs today, but they nonetheless indicate a real unease that continues to influence us, often quite unconsciously. For example, the fact that we avoid speaking ill of the dead, remembering only the positive aspects of their lives, is no doubt largely rooted in a social convention stemming from the fear they arouse in us.

Ghoulish rebirths

Vampires, werewolves and other creatures from beyond the grave are a good illustration of the unease that death causes us and of the need to transcend this fear by creating mythical characters that exist on the fringes of the living world. Even though these myths do not have the same influence as they once did and now owe their survival to their roles in horror books or movies, their origins are nonetheless fascinating. They show just how much human beings need the supernatural to overcome their incomprehension and distress in the face of a phenomenon as natural as death.

> Engraving portraying a werewolf devouring a young woman.

A thirst for blood

A classic symbol of life, blood has always held an essential place in a number of religious rituals. To purify themselves, the Persians sprinkled themselves with the blood of an animal sacrificed during the rites that were part of the cult of the god Mithra. During the same period, the Greeks drank animal blood mixed with wine during rites celebrating Dionysus. As for the Aztecs, they were truly obsessed with the power of blood, even believing that the sun needed this precious liquid as a source of energy to move around the heavens. Indeed, blood flowed like water in pre-Columbian Mexico; several thousand people had their throats slit to satisfy the appetites of the gods (as well as the emperors!) and to please the living, who believed they would in return be blessed with fertility and immortality. Even though such sacrifices are completely unimaginable today, the symbolism of blood as the bearer of life remains anchored in tradition, as shown by the importance of the transubstantiation of bread and wine into the flesh and blood of Christ in the Christian Eucharist.

Symbolism aside, the legend of vampires, immortal beings who draw their life force from the blood of others, seems to have a biochemical origin. Indeed, certain physiological disorders that are genetic in origin display a number of the characteristics ascribed to vampires, in particular a heightened sensitivity to the sun. These disorders, called porphyrias, are caused by a deficiency in certain enzymes involved in the production of heme, the pigment that causes iron to bind with hemoglobin. Without these enzymes, pigments called porphyrins accumulate in abnormally high quantities in the body's organs, especially in the liver, bone marrow and skin. Porphyrins are purplish-red pigments that absorb UV rays from light and produce free radicals that can cause enormous damage to the tissues and lead to the appearance of a number

^ Aztec ritual of human sacrifice
(*Codex Magliabechiano*)

of characteristics associated with vampires. In late-stage skin porphyria, an enzyme deficiency (uroporphyrinogen decarboxylase) results in the accumulation of uroporphyrins, molecules whose fluorescent properties damage the skin and give the teeth and nails a reddish color following sun exposure. The effects of congenital erythropoietic porphyria, a very serious illness caused by the absence of the enzyme uruporphyrinogen cosynthase, are even more dramatic: the excess porphyrins cause lesions not just on the skin, but also in several tissues, among them the gums, which make the teeth appear to protrude, giving them a fang-like appearance. In this kind of porphyria, an anomaly in the hemoglobin leads to the elimination of red blood cells and serious anemia. In other words, some people suffering from porphyrias have very pale skin and reddish fangs and have to avoid sunlight as much as possible. Clearly, the resemblance to fictitious vampires is more than a little disturbing! Not to mention that porphyrias may also cause abnormal hair growth (hypertrichosis) all over the body, a hairiness that may also have played a role in the development of werewolf legends.

The role played by these natural phenomena in the origins of vampire and werewolf myths remains unclear, but we do know that these living dead have their equivalent in a great many cultures in Europe, Africa, the Middle East and Asia. Ch'ing Shih in ancient China, Kyuuketsuki in Japan, Pennaggalan in Malaysia, Kali in India and, of course, the many vampires of Eastern Europe (the upyr and strigoi, for example)—all these creatures could on occasion reappear among the living and gorge themselves on human blood to regain their strength.

Zombies

In the Voodoo religion, a zombie is a dead person taken from the grave and held in a state of slavery by the magic of a *bokor* (wizard). According to legend, the wizard's victim is first struck down by a *coup de poudre* (a hit of powder), a potion that makes physiological functions slow down to such an extent that the person appears to be dead. After retrieving the body of the person who was buried alive, the *bokor* gives his victim a

> Men dressed as zombies for a voodoo ritual in Haiti.

Fishy poison

Puffer fish (*Tetraodon* sp.) owe their name to their strange ability to fill their stomachs with water and puff up when they feel threatened. In addition to this defensive performance, all puffer fish contain tetrodotoxin, an extremely toxic molecule that prevents sodium from entering neurons and causes total muscular paralysis by irreversibly blocking the transmission of nerve impulses. Tetrodotoxin is not produced by the fish themselves, but rather by certain bacteria found in the plants on which they feed. Thanks to a mutation in the structure of their sodium channel, *Tetraodon* species are completely immune to this toxin, and it accumulates in their livers and reproductive organs without causing any harm. This symbiosis is particularly beneficial to the fish, as their high toxin content makes them impossible for predators to eat!

Very widespread in the world's oceans, puffer fish are particularly popular in Japan, where a genus of this family, the *Takifugu* (better known by the name "fugu"), holds a special place in that country's culinary tradition. The most popular (and the most dangerous) is the torafugu (*Takifugu rubripes*). Each fish contains enough toxin to kill 30 adults. Fugu consumption is strictly regulated, and only chefs with special training in preparing the fish are authorized to serve it, usually in the form of sashimi sliced so thinly that the pattern on the serving platter remains visible. Some especially skilled chefs prepare the fish in such a way that it contains traces of toxin and causes tingling and numbness in the tongue and lips.

Several thousand times more powerful than cyanide, this toxin does not enter the brain, with the result that it causes a horrible death by asphyxiation; the person remains fully aware of the paralysis that gradually overtakes the entire body. In some cases, victims of fugu who appeared to be dead have recovered a few days later, just before their body was to be cremated! To avoid this sort of premature funeral, the inhabitants in certain parts of Japan would leave the victim beside his coffin for 3 days before holding the funeral.

< A fugu (*Cyclithys spilostylus*) in its natural habitat.

^ A platter of fugu

second potion that keeps him in a state of living death, with no soul of his own and completely subjugated. Some ethnobotanists have suggested that the profound lethargy induced by the *coup de poudre* might be caused by a mixture of tetrodotoxin, a powerful toxin derived from a kind of puffer fish (*Sphoeroides maculatus*), and bufotoxin, which is extracted from the cane toad (*Rhinella marina*). This is a dangerous cocktail to say the least. Tetrodotoxin blocks the transmission of nerve impulses by neurons, which causes muscle functions to cease, as well as a loss of control of basic functions such as the regulation of blood pressure and temperature (see box on p. 87). In addition, certain bufotoxins, such as bufotenin, have a structure similar to that of psilocin (the active molecule in magic mushrooms), giving it hallucinogenic properties. As for the second potion used to enslave the living dead, it has been suggested that it might be made from the thorn apple (*Datura stramonium*), an extremely toxic nightshade containing an alkaloid (scopolamine) known to cause nightmarish hallucinations, amnesia and loss of consciousness. Even though it is unlikely that these poisons are the only ones responsible for the emergence of the zombie myth and its influence on Voodoo culture over the centuries, it is nonetheless interesting that *Datura stramonium* is called the Zombie cucumber in Haiti and that using these kinds of poison for "zombification" is expressly forbidden by the Haitian penal code:

> Art. 246. Is considered to be poisoning, any attempt on the life of a person, using substances that can cause more or less immediate death, no matter how these substances have been used or administered and whatever their consequences. (Penal code 240, 247, 262, 263, 334, 372.)
>
> Is also considered to be an attempt on the life of a person by poisoning, any use against a person of substances that, without killing the person, have produced a lethargic state of greater or lesser duration, no matter how these substances were used and whatever their consequences.
>
> If, as a result of this lethargic state, the person was buried, the attempt will be considered murder. (Penal code 241 and following Act of 27 October, 1864, as amended.)

Just as in nature, where life always follows death, human death is interpreted as a passage into another world and the beginning of a new life. Our attitude is therefore ambivalent, a mixture of fear, hope and fascination. The fear of death, as we have stressed, is clearly biological, written indelibly in our genes, and dictates the avoidance and fighting behaviors necessary for our survival. Hope and fascination with death, on the other hand, are typically human behaviors that are more a reflection of our brain's inability to accept that the death of our individual selves is inevitable.

> Japanese demon mask

Chapter 4

The Wear and Tear of Time

It's all very well to have an iron constitution—
we always rust out in the end.
Jacques Prévert (1900–1977)

The dramatic improvement in sanitary conditions, nutrition and the treatment of infectious diseases during the 20th century has had extraordinary repercussions for people's life expectancy. Whereas in 1900 barely 1% of the world's population was 65 or over, this proportion reached 10% in 2000 and could even go as high as 20% in 2050, with no fewer than 2 billion "old people" living on the planet at that time. Right now, the average person on earth is under 30; by the time this person dies, the average age will be 50. For the first time in human history, the person representing the majority of people will no longer be a young adult but a graying individual with a slightly wrinkled face.

Population aging of this magnitude brings with it many social changes, the biggest being without a doubt a significant deterioration in quality of life for a large number of people living to an age that was not reached in the past. Just because you live longer, it doesn't necessarily mean you do so in good health. Quite the contrary—increased longevity is unfortunately often associated with a parallel increase in the incidence of many chronic diseases that counteract the advantages a longer life might have. In these conditions, reaching an advanced age does not really offer any advantages, especially when uncertain health causes a loss of self-sufficiency and much suffering, both physical and psychological. Old age then becomes an ordeal, a painful journey to the end of life that you have to resign yourself to

< Albrecht Durer, *Study of a Man, Aged Ninety-Three*

enduring while awaiting death's deliverance. No one wants to die like this, and, for many, a large part of the fear associated with death does not stem from the end of life in itself but rather from the dread of experiencing this kind of loss of quality of life before dying.

Old age and diseases must not, however, be confused with each other. Even though the incidence of most of the chronic diseases that affect us increases significantly with age, it is nonetheless perfectly possible to grow old in good health, and die of natural causes, without losing many enjoyable years and experiencing the slow death often associated with chronic illnesses. An impressive number of studies carried out in recent decades show that the lifestyle typical of industrialized societies, especially our diet, obesity and sedentary habits, plays a major role in the development of these diseases and in the loss of quality of life. In practice, it is entirely possible to grow old in good health by adopting a few key lifestyle habits that considerably reduce the risk of falling ill with a chronic disease (see Figure 1).

The seeming inevitability of fatal diseases is very often a convenient excuse for people who do not want to change their lifestyle habits. When they are faced with the challenge posed by these kinds of changes, it is not uncommon to hear that it isn't worth the trouble, since "you have to die from something." The reality, however, is different. Chronic diseases are not unavoidable; on the contrary, it is entirely possible to remain in good health, physically, intellectually and emotionally, until very late in life (see Figure 2). At that point, the wearing effects of time on vital functions causes the organism's equilibrium to deteriorate rapidly, leading to what is often a rapid death from natural wear and tear or from the inability to fight off an attack from microorganisms (pneumonia is a frequent cause of death in very elderly people). While it may seem paradoxical, adopting healthy lifestyle habits so as to reduce the risk of becoming ill with a debilitating disease is not only a way to improve life expectancy and quality, but it is also the best way to die with the greatest dignity possible.

The Five Golden Rules to Prevent Chronic Illness

1. Don't smoke

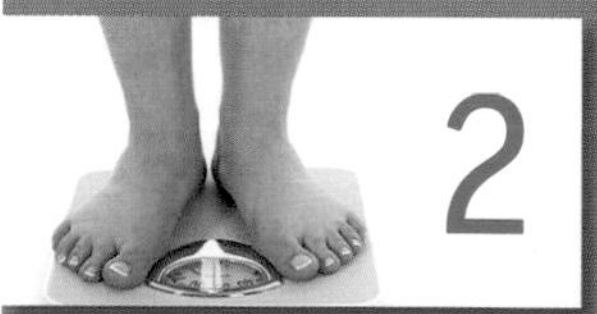

2. Maintain a normal weight (BMI between 19 and 24)

3. Eat plenty of plant foods—fruits and vegetables and whole grains

4. Be physically active at least 30 minutes every day

5. Reduce consumption of sweet, fatty and salty foods, especially fast foods

Percentage of chronic diseases that can be prevented by these five lifestyle changes	
Type 2 Diabetes	90 %
Heart disease	82 %
Cancer	70 %
Strokes	70 %

Figure 1

Modern-day Methuselahs

Although life expectancy at birth made a dramatic leap forward in the last century, rising from 47 to nearly 80 (85 for women), most of this increase is linked to a significant decline in infant mortality and mortality due to infectious diseases. In fact, the number of very old people (90 or over) is growing much more slowly and remains even now an extremely rare phenomenon: it is estimated that just one person in 10,000, or .01% of the population, lives to be 100

Figure 2

< Jeanne Calment, to date the longest-lived person (122 years, 164 days)

and, among these centenarians, only 1 in 1,000 lives to celebrate a 110th birthday. The (known) record for longevity belongs to Jeanne Calment, a Frenchwoman who died at the venerable age of 122 years and 164 days. Very active until she was 110 (she was still riding a bicycle at 100!), Calment remained in good health until a month before she died. Her main problems were her fragile bones, as well as the gradual loss of her hearing and sight. Such long life spans are obviously exceptional. Nevertheless, they provide a striking example of the human body's potential for maintaining vital functions up to a very advanced age, when chronic diseases are avoided or at least postponed. Furthermore, it is interesting to note that at certain very advanced ages (95 and over), the mortality rate becomes much lower than generally expected (see Figure 3). This drop is associated with a declining risk for some diseases (notably cancer) at these ages.

Autopsies carried out on supercentenarians (aged 110 and over) have shown that their deaths were caused not by the typical diseases of old age (cancer, cardiovascular disease, Alzheimer's), but rather by a degeneration of the heart caused by the accumulation of certain protein deposits that, over time, eventually block the vessels irrigating the organ. Sort of the way water pipes in

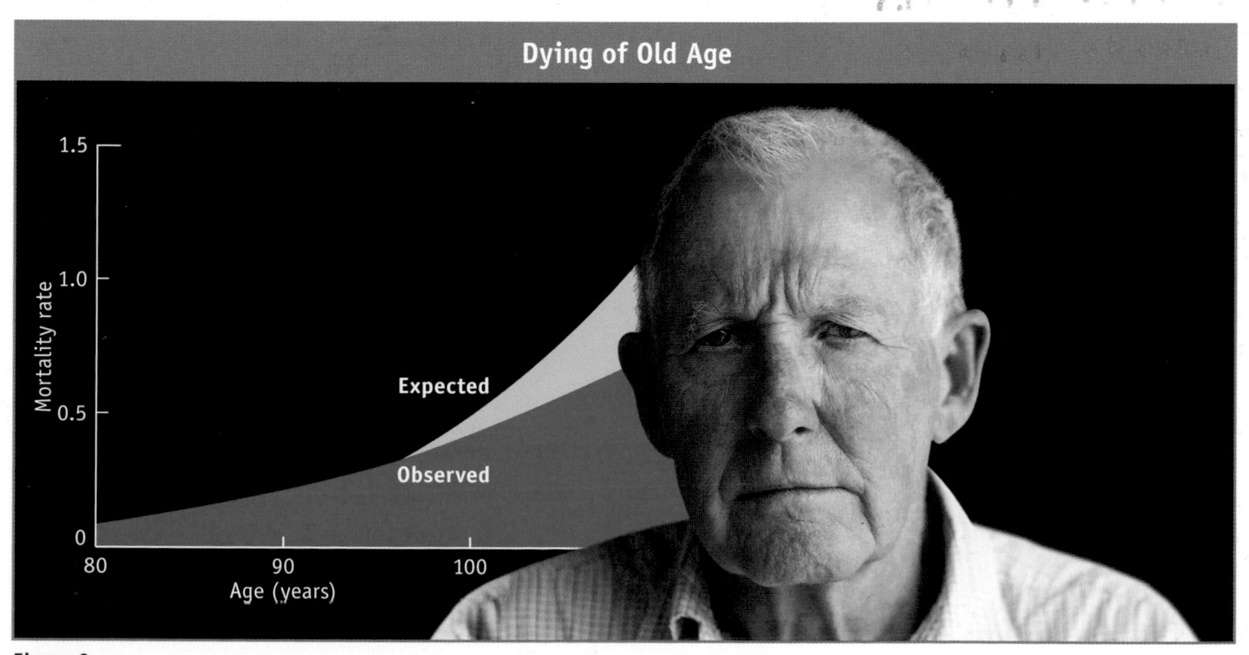

Figure 3

an old house eventually wear out! Growing old in good health therefore means postponing the onset of diseases for as long as possible, so that the loss in quality of life they cause is limited to the shortest period possible. In other words, centenarians are striking proof that a natural death is very often the ending to a healthy life.

A radical process

From a biological standpoint, aging is the result of a compromise made by organisms that must choose between the need to survive long enough to reproduce and the considerable expenditure of energy needed to defend themselves against the never-ending attacks they are subjected to. Giving priority to defence mechanisms designed to resist these attacks would require too large an investment of energy to allow for efficient reproduction. On the other hand, surviving merely with the purpose of reproducing, which is what simple organisms like bacteria do, means retaining as simple a genetic and cellular structure as possible and, consequently, hinders the development of complex organisms.

Aging is the result of a gradual accumulation throughout life of all kinds of damage, on both the cellular and molecular levels (see Figure 4), resulting in biological systems that become weaker and function poorly, leading ultimately to death.

Even though this damage occurs randomly and is inevitable in human life, there are nonethe-

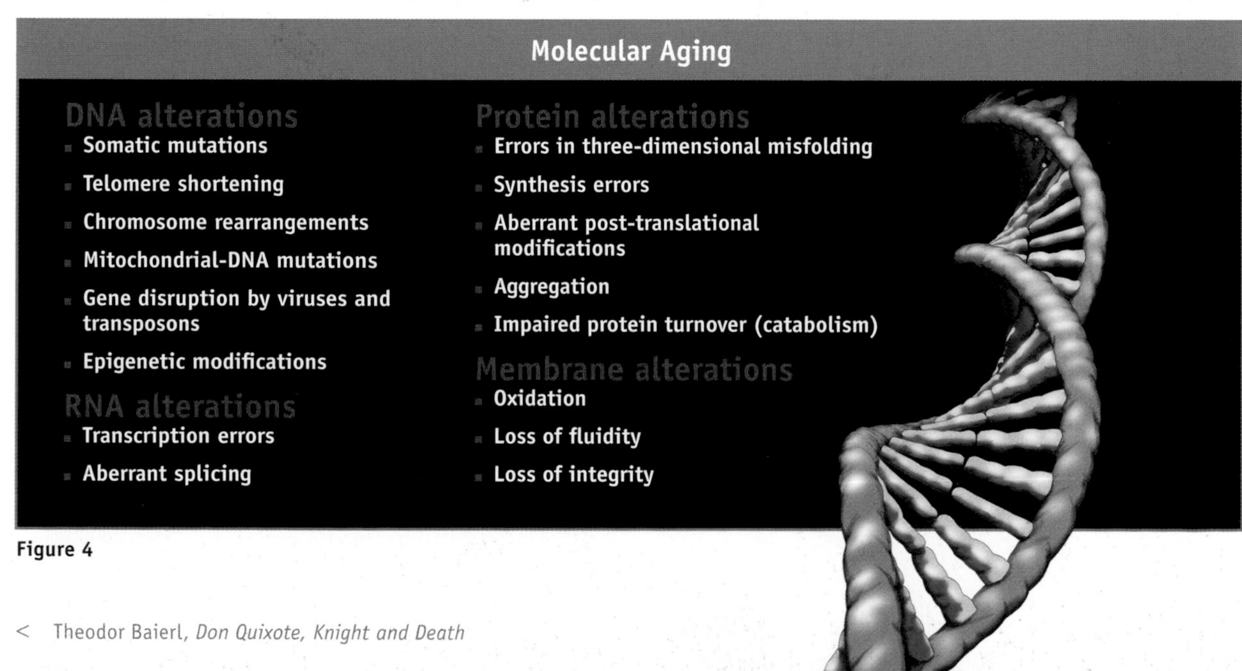

Figure 4

< Theodor Baierl, *Don Quixote, Knight and Death*

less several factors that determine the speed with which it can alter cellular functions. The first factor is related to our built-in ability to repair damage to our genetic material or to other cell components using certain defense systems that are part of our cellular makeup. These hereditary protection mechanisms are the key players in the genetic component of longevity: it is estimated that about one-third of the people who reach an advanced age owe it to a lucky draw in the human species' gene pool lottery. Members of some families, for example, live much longer

^ Norman Vaughan, 88, scales the Antarctic summit of the mountain that now bears his name.

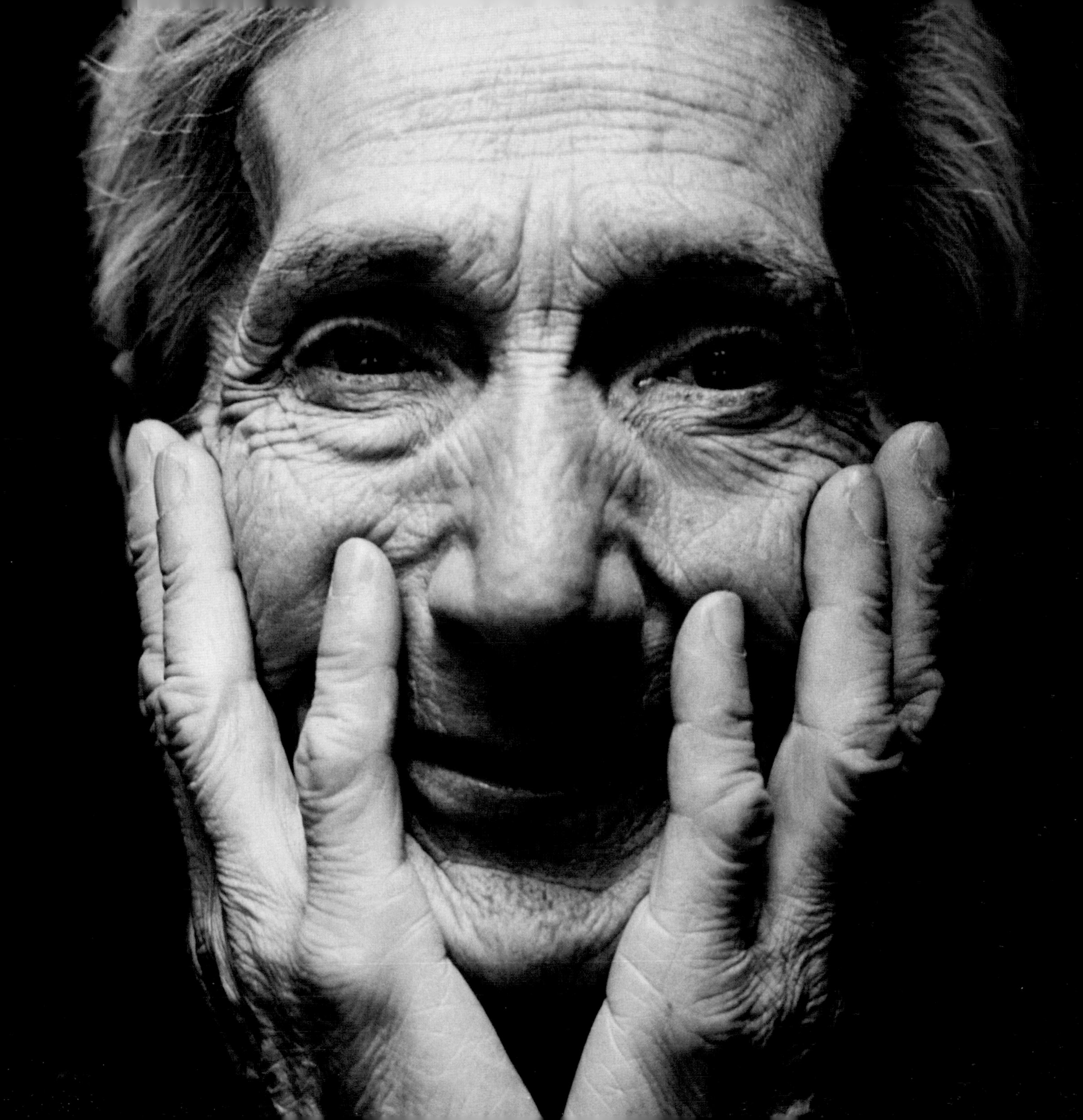

than the general population. Thus, the sibling of a centenarian is 17 times likelier to also reach 100. Studies have shown that, in a good many of these families, the types of enzymes involved in DNA repair are more active. In other families, it is the genes involved in transporting and metabolizing HDL cholesterol that seem to be involved. Conversely, some unfortunate people inherit deficient types of certain DNA-repairing enzymes and age prematurely. In Werner's syndrome, for example, a single mutation in a gene essential to maintaining DNA integrity causes the aging process to speed up dramatically, with the result that, beginning in early adulthood, people with this syndrome display characteristics usually observed in much older people (hair loss, cataracts, etc.). These people often die before reaching their 50s, usually as a result of cancer or cardiovascular disease.

The importance of these protective systems is due to the constant attack the components of our cells undergo from oxygen byproducts produced by the normal functioning of our organs. Even though it is essential for life, oxygen nonetheless remains an unpredictable substance that can react with a number of molecules when it is activated. Most of the time, the product of these reactions is positive. For example, when cells absorb glucose or fats, they require oxygen to convert the energy contained in these molecules into ATP (see Chapter 2). However, the mechanism involved in this transformation is not perfect, and a certain number of "waste products," called free radicals, form at the same time (see Figure 5). These byproducts have a powerful ability to oxidize and can attack nearby structures, in a process that is analogous to the corrosion that causes rust to form on metal. Even though most of these free radicals are changed into totally harmless molecules, thanks to our protective systems, notably superoxide dismutase (SOD), some of them manage to escape their vigilance and damage our genetic material in particular. This phenomenon is far from insignificant, since it is estimated that, every day, our DNA is subject to 10,000 attacks from free radicals. In other words, we are rusting from the inside out!

Some experimental data show that aging and many diseases that come with age (cancer, cardiovascular diseases, Alzheimer's) are caused in part by these sustained attacks from free radicals and that reducing their toxic effects is essential to living a long and healthy life. The discovery that plant foods contain large quantities of antioxidants, molecules that have the ability to neutralize the effects of free radicals, suggests that the positive effects of these foods in reducing chronic diseases are due in part to these antioxidant properties.

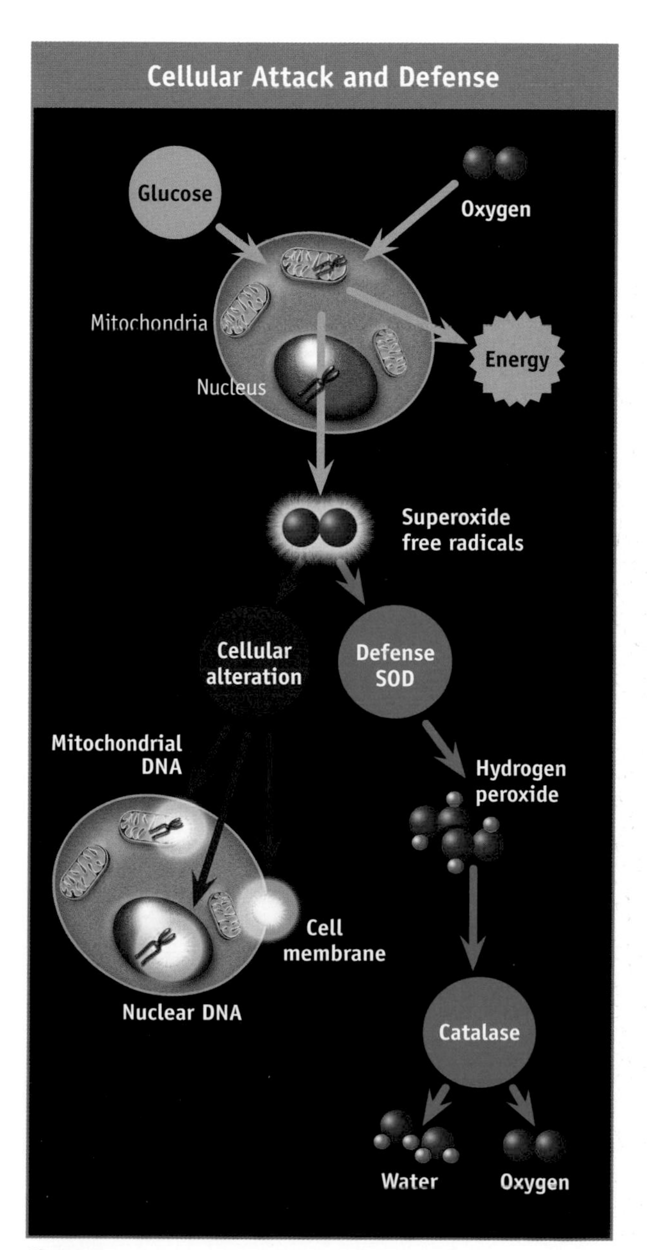

Figure 5

Caloric restriction

It was been known for more than a century that animals consuming fewer calories (as long as they do not have nutritional deficiencies in essential elements) live much longer than those who ingest larger quantities of food. In mice, for example, a 30% reduction in caloric intake makes them live 40% longer, due in large part to a significant reduction in cardiovascular disease, cancers and neurodegenerative diseases. Similar effects of caloric restriction have been observed over and over again in many animals (worms, flies and fish) and more recently even in primates. The latter result is especially interesting given the close evolutionary relationship these animals have with the human species. Monkeys subjected to caloric intake restriction are noticeably more lively and have more elastic skin and excellent blood lipid and blood sugar profiles compared with those who eat more. The latter show some of the typical signs of aging, especially hair loss, the onset of wrinkles and elevated levels of blood lipids and sugar.

Caloric restriction naturally leads to weight loss, but its beneficial effects go far beyond the advantages that maintaining a healthy weight may offer. It is likely that the dramatic improvement in quality of life and life expectancy that results from reducing caloric intake is associated with decreased free radical production. Indeed, less food means that the mitochondria use less oxygen and transform energy into ATP more efficiently, which in both cases means that fewer free radicals are produced.

However, we now know that the impacts of caloric restriction on longevity are much more complex, since it also activates a number of defense systems involved in the stress response, mainly those involving a class of proteins called sirtuins. The activation of these enzymes leads to a whole range of positive effects that together

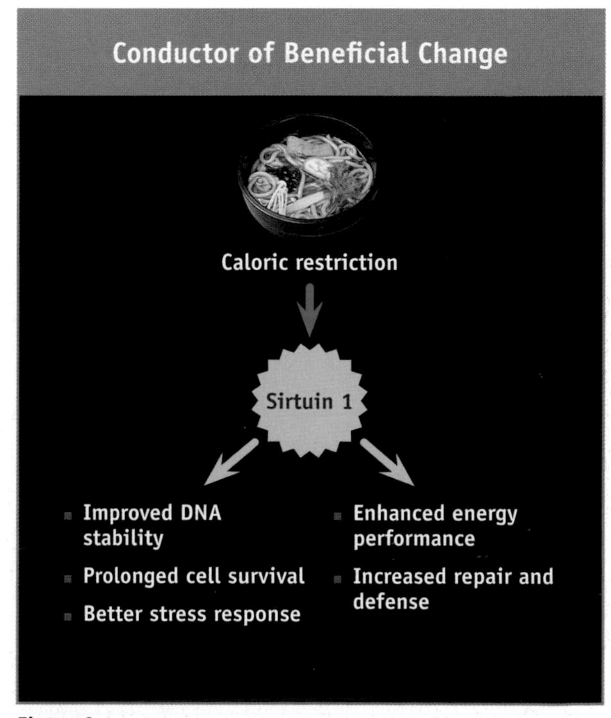

Figure 6

reduce cellular aging (see Figure 6). In particular, these enzymes enable certain especially vulnerable areas of DNA to adopt a more compact organization that reduces their sensitivity to external attacks.

One of the most exciting discoveries in research on aging is the observation of certain molecules that are able to activate these survival mechanisms and thus imitate the effects of caloric restriction. For example, the activation of certain sirtuins by resveratrol, a polyphenol found in red wine, causes a remarkable increase in longevity in simple organisms like yeast, worms and even some species of fish. The creation of molecules similar to resveratrol, and therefore able to increase life expectancy, is currently a very active area of research. If this research should be successful, a molecular "fountain of youth" of this kind could have extraordinary impacts on the longevity of the human species.

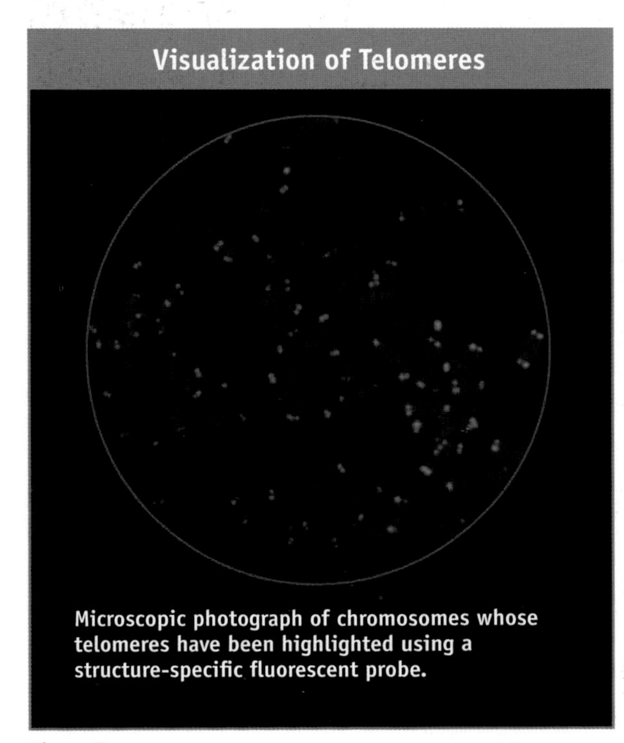

Microscopic photograph of chromosomes whose telomeres have been highlighted using a structure-specific fluorescent probe.

Figure 7

Costly losses

Throughout our lives, the cells that make up our bodies have to constantly renew themselves to keep our organs functioning well (see Chapter 2). This is an incredibly complex process in which the roughly 3 billion components (nucleotides) of our genetic material (DNA), which are organized into 23 distinct pairs of chromosomes, must be faithfully copied for transmission to daughter cells. This mechanism works very well on the whole, but it does have an inherent "functional defect": it is unable to copy the DNA located at each end of the chromosomes, regions called telomeres (see Figure 7). As a result, each time a cell copies its genetic material so it can divide, the ends of the chromosomes are irretrievably lost (see Figure 8). Unfortunately, these telomeres keep getting shorter over time and finally reach a critical length, and the cell becomes unable to

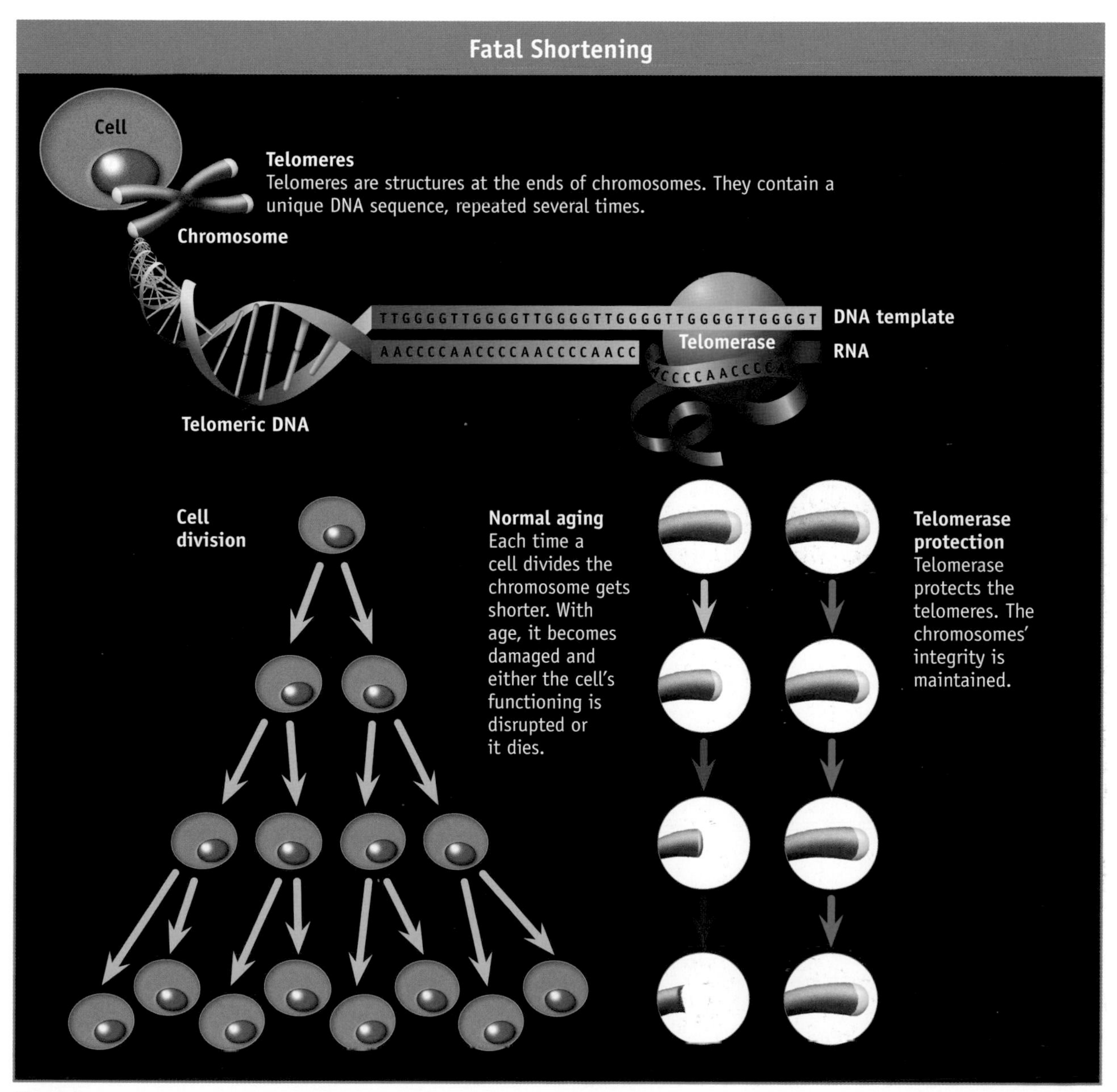
Fatal Shortening
Cell
Telomeres
Telomeres are structures at the ends of chromosomes. They contain a unique DNA sequence, repeated several times.
Chromosome
TTGGGGTTGGGGTTGGGGTTGGGGTTGGGGTTGGGGT
DNA template
AACCCCAACCCCAACCCCAACC
Telomerase
RNA
Telomeric DNA
Cell division
Normal aging
Each time a cell divides the chromosome gets shorter. With age, it becomes damaged and either the cell's functioning is disrupted or it dies.
Telomerase protection
Telomerase protects the telomeres. The chromosomes' integrity is maintained.

Figure 8

regenerate and dies. The gradual loss of telomeres is a key factor in the aging of our organism and in our life expectancy.

The importance of telomeres is clearly shown in the development of cancer. Whereas normal cells have a limited life span, one of the basic characteristics of cancer cells is their immortality, that is, their ability to reproduce indefinitely. One way they stay "immortal" is by putting in place systems to interfere with telomere loss: in the vast majority of cancers, tumorous cells synthesize telomerase, an enzyme that can regenerate telomeres that have not been copied by the replication machinery. Because of this enzyme, cancerous cells manage to maintain their chromosomal integrity even when they divide at a frenzied pace (see Figure 8).

While telomere loss seems inevitable in normal cells, this phenomenon may nonetheless be significantly modified by certain lifestyle factors. For example, recent studies suggest that adopting an active lifestyle and combining with a diet rich in plant foods slows down the loss of telomeres while significantly increasing telomerase activity. Although aging is an inexorable process, it is nonetheless possible to slow it down by making the most of the health capital nature has given us, thus avoiding debilitating chronic diseases. It really is possible to die of old age, to reach the stage where the wear and tear of time has taken its toll and the incredible flow of energy that is key to maintaining vital functions dries up and becomes insufficient to conduct the magnificent symphony we call life. Even though no one likes the prospect of dying, coming to the end of life at an advanced age after a very full existence is much easier to accept. It is really the prospect of dying prematurely and of suffering that terrifies us, a fear often associated with chronic diseases.

> Laurits Andersen Ring, *Old Man Walking in a Rye Field* (detail)

Chapter 5

Dying Little by Little

I have no terror of Death.
It is the coming of Death that terrifies me.
The Picture of Dorian Gray (1890), Oscar Wilde (1854–1900)

Good health requires coordinated action from all of the body's organs, with each one having to constantly listen to the others' needs so as to maintain an optimal equilibrium able to adequately support the organism's functions. However, the mechanical process that maintains health is not perfect, as shown by the considerable number of disorders that can affect any of our organs and cause serious diseases that put our lives in danger. To date, 3,600 diseases affecting the human body's physiological systems have been identified by the World Health Organization (WHO). Whether they are hereditary, linked to lifestyle or caused by external factors (accidents and others), their vast diversity shows the scope of the challenge facing the medical community in their attempts to improve the health of the population (see Figure 1).

In the United States and Canada, just as in most other industrialized countries, the principal causes of mortality lie in the high incidence of chronic diseases. Cancer, cardiovascular diseases, lung diseases, diabetes and neurodegenerative diseases like Alzheimer's disease alone are responsible for more than two-thirds of the deaths recorded each year (see Figure 2).

Sadly, these diseases too often strike prematurely, depriving people of several years of good health. As mentioned in the previous chapter, it is possible to postpone the onset of these diseases by adopting healthy lifestyle habits, in particular by avoiding smoking, paying special

< Detail from the *Pietà* by Michelangelo

attention to eating habits and getting regular physical activity to avoid becoming overweight. These precautions obviously cannot completely eliminate the risk of getting diseases, especially at an advanced age. However, by enabling us to significantly postpone their onset, a preventive approach can considerably enhance quality of life while reducing the length of the period of disease and suffering that often precedes the end of life.

Chronic diseases also require a complex and expensive treatment strategy, which are only available in a hospital setting. Aside from the enormous pressure they place on the funding and functioning of health systems, the high incidence of these diseases has major ramifications for the very process of dying. Whereas in the past it was a personal ordeal, experienced in the privacy of the family circle, over 80% of deaths now occur in hospitals, resulting in a medicalization of death that completely redefines the social context in which the final moments of life are spent. Mortality caused by chronic diseases can therefore be considered to be the modern version of

International Classification of Diseases

- Infectious and parasitic diseases
- Neoplasms
- Diseases of the blood and blood-forming organs and disorders involving the immune mechanism
- Endocrine, nutritional and metabolic diseases
- Mental and behavioral disorders
- Diseases of the nervous system
- Diseases of the eye and adnexa
- Diseases of the ear and mastoid process
- Diseases of the circulatory system
- Diseases of the respiratory system
- Diseases of the digestive system
- Diseases of the skin and subcutaneous tissue
- Diseases of the musculoskeletal system and connective tissue
- Diseases of the genitourinary system
- Pregnancy, childbirth and the puerperium
- Certain conditions originating in the perinatal period
- Congenital malformations, deformations and chromosomal abnormalities
- Symptoms, signs and abnormal clinical and laboratory findings, not elsewhere classified
- Injury, poisoning and certain other consequences of external causes
- External causes of morbidity and mortality
- Factors influencing health status and contact with health services

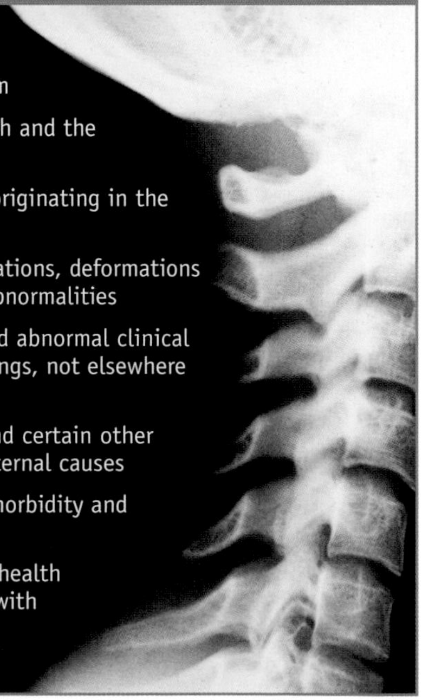

Figure 1

Source: http://en.wikipedia.org/wiki/ICD-10 - List

death, whose mechanisms we need to understand if we want to better come to grips with the factors that cause death.

Problematic circulation

Dying from a heart attack or stroke (cerebrovascular accident, or CVA) is often considered to be the perfect example of a "good death," a fatal event that strikes suddenly and causes rapid death, without interminable suffering. There is some truth to this view, for these two events can indeed be extremely efficient killers owing to their ability to instantly block the supply of oxygen to the two essential organs that dependent on this precious gas the most: the heart and the brain.

Despite their often sudden nature, these deaths are generally the end result of a long process of blood vessel deterioration, during which the gradual accumulation of cholesterol and various other substances causes the formation of atheromatous plaques that grad-

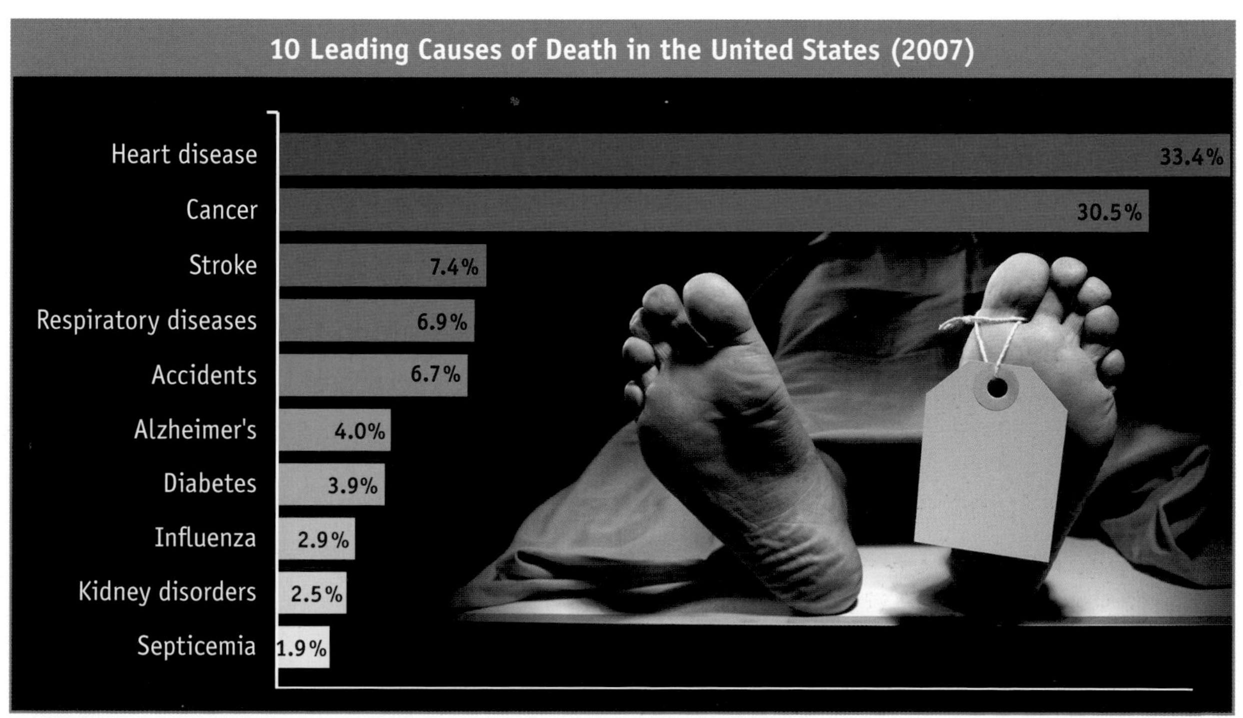

Figure 2

Source: Centers for Disease Control

ually reduce the blood's ability to flow to its destination (see Figure 3).

When these plaques break off, the lesion that forms on the wall of the blood vessel is interpreted by our body's defense systems as a wound to be healed. This causes a blood clot to form, which can completely block the vessel, keeping oxygen from reaching the target organs. Heart attacks and strokes are therefore called ischemic diseases (from the Greek *ischein*, meaning to restrain, and *haima*, meaning blood) and result from a strangulation of the vessels, causing sudden oxygen deficiency.

Myocardial infarction

Even though its incidence has greatly decreased in recent decades thanks to prevention and improvements in therapeutic intervention, heart attacks (myocardial infarction), remain the most common cardiovascular event and are often the most deadly: approximately a third of heart attack patients die from the attack, usually within the hours immediately following its onset. This fatal outcome is due to the fact that the vessels blocked by the breaking off of atheromatous plaques are coronary arteries, a series of vessels

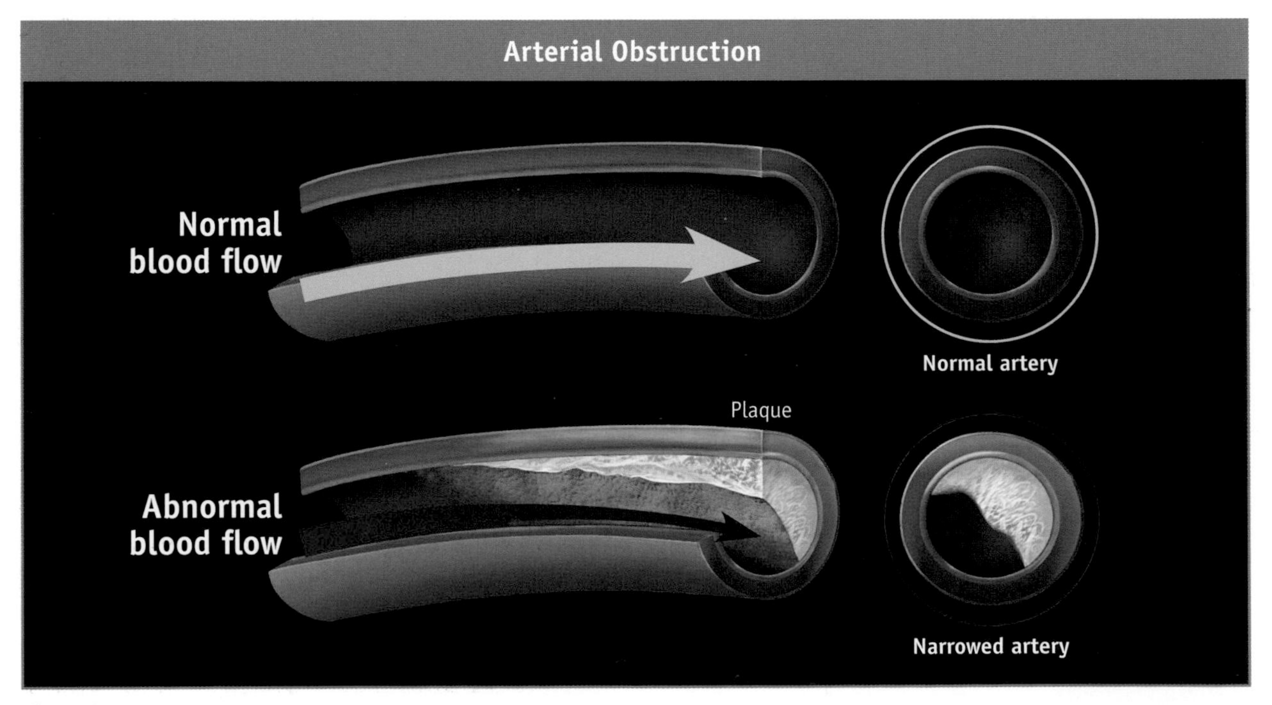

Figure 3

Source: www.pdrhealth.com

whose role is to carry oxygen and nutrients to the heart muscle cells. Muscle cells deprived of oxygen die, as they can no longer perform the contractions needed for the heart to function. Depending on the extent of the damage caused by oxygen deprivation, death can occur quickly once the blood circulation stops (see Figure 4).

These sudden deaths are very often the first sign of a heart problem, as most people affected show no symptoms that would even hint at such an abrupt ending to their lives. In young people (under 35) in good physical health, sudden cardiac deaths are mainly due to congenital disorders such as hypertrophic cardiomyopathies, diseases characterized by changes in the structure of heart muscle tissue. In older people, however, the overwhelming majority of these sudden deaths can be prevented, since they are the consequence of coronary diseases whose development is directly linked to lifestyle, in particular smoking, diet and physical inactivity.

While these sudden deaths can occur at any time, intense emotions are the most dramatic trigger of cardiac malfunction. Anger, fear or a powerful emotion (positive or negative) cause excessive stimulation of the adrenergic nervous system and can result in an abnormally rapid heart rate (ventricular tachycardia) or chaotic heart contraction (ventricular fibrillation), with all of these factors considerably increasing the risk of sudden death. The impacts of severe emotional stress are well illustrated by the significant increase in sudden deaths during tragic events like natural disasters. For example, an analysis of the deaths that occurred in the hours immediately following the violent earthquake that shook southern California in January 1994 showed a five-fold increase in the number of people who died suddenly.

A number of aspects of emotional stress are linked to heart diseases. In left ventricular malfunction, which occurs particularly in older women, a traumatic or emotionally stressful experience leads to an increase in catecholamines, causing sudden pain in the chest and

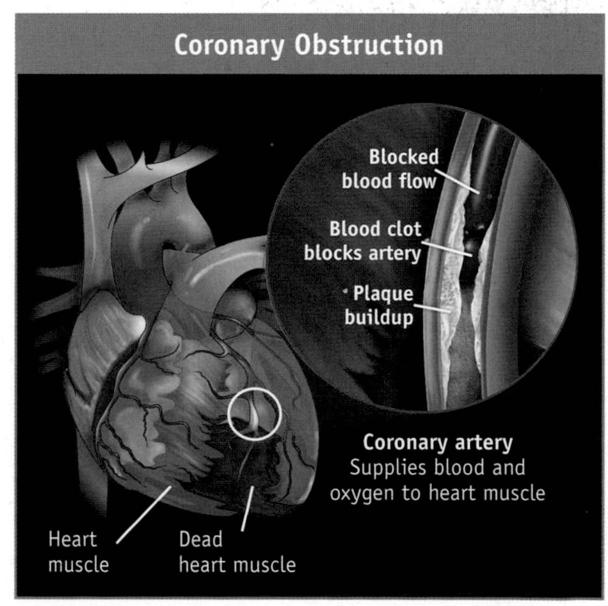

Figure 4 Source: www.pdrhealth.com

lack of breath. In the case of myocardial ischemia, following intense stress some people may have ischemic episodes, even if clinical and stress tests are negative. In the same way, intense emotions may cause ventricular arrhythmias in approximately 20% of clinical cases.

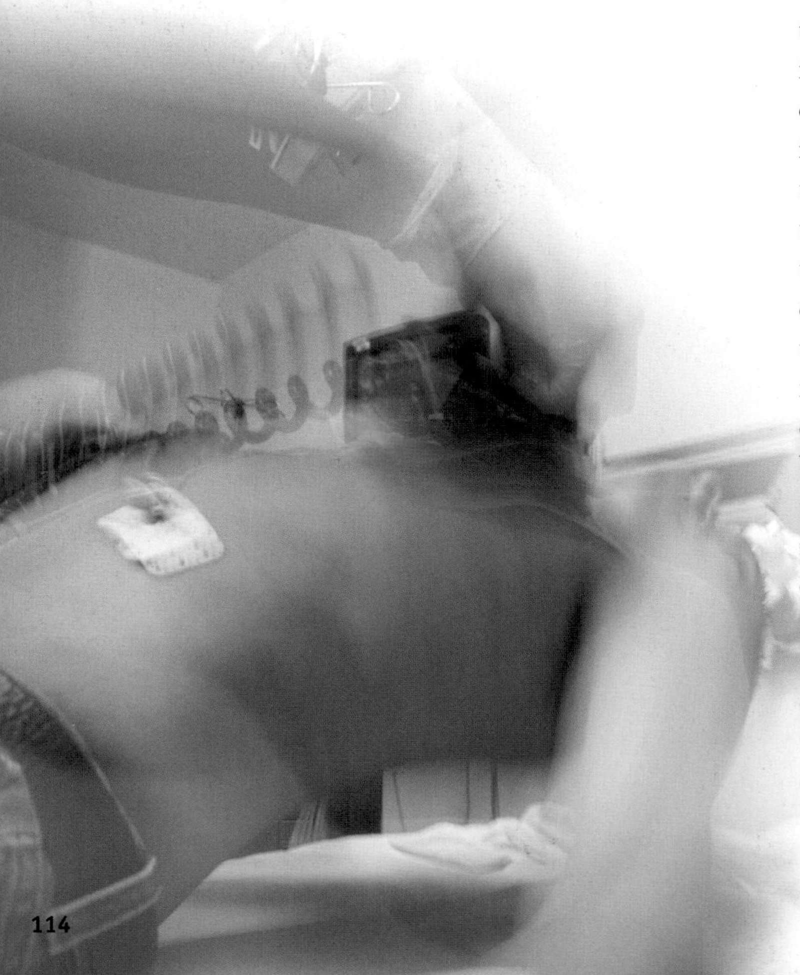

Cerebrovascular accidents (strokes)

The third most common cause of death in industrialized countries, cerebrovascular accidents (CVAs), commonly called strokes, are caused by a sudden interruption in blood circulation in the brain as the result of a blockage or the rupture of a vessel irrigating brain cells. As with the heart, this event is often fatal, since a constant blood supply is essential for neurons to be able to receive the oxygen and nutrients their normal functioning depends on: just a few minutes after circulation stops, these cells begin to undergo irreparable damage and quickly lose their ability to transmit a nerve signal. The brain is not just the center of thought and intelligence; it is also the command center that controls all motor processes, both voluntary (e.g., speech) and involuntary (e.g., breathing). This is why one of the first symptoms of stroke is numbness–a loss of feeling or function in certain parts of the body. These effects are attributable to the fact that nerve impulses are no longer being transmitted. If the part of the brain affected by the deficiency is essential to a basic physiological function, the control of breathing, for example, the consequences are deadly and the person may die quickly. When the blockage affects parts of the brain that are less crucial to survival, people can survive but usually with a significant loss

< A cardiac defibrillator being used on a patient.

< Electron microscope image of red blood cells (red) trapped within fibrin (gray) to form a blood clot.

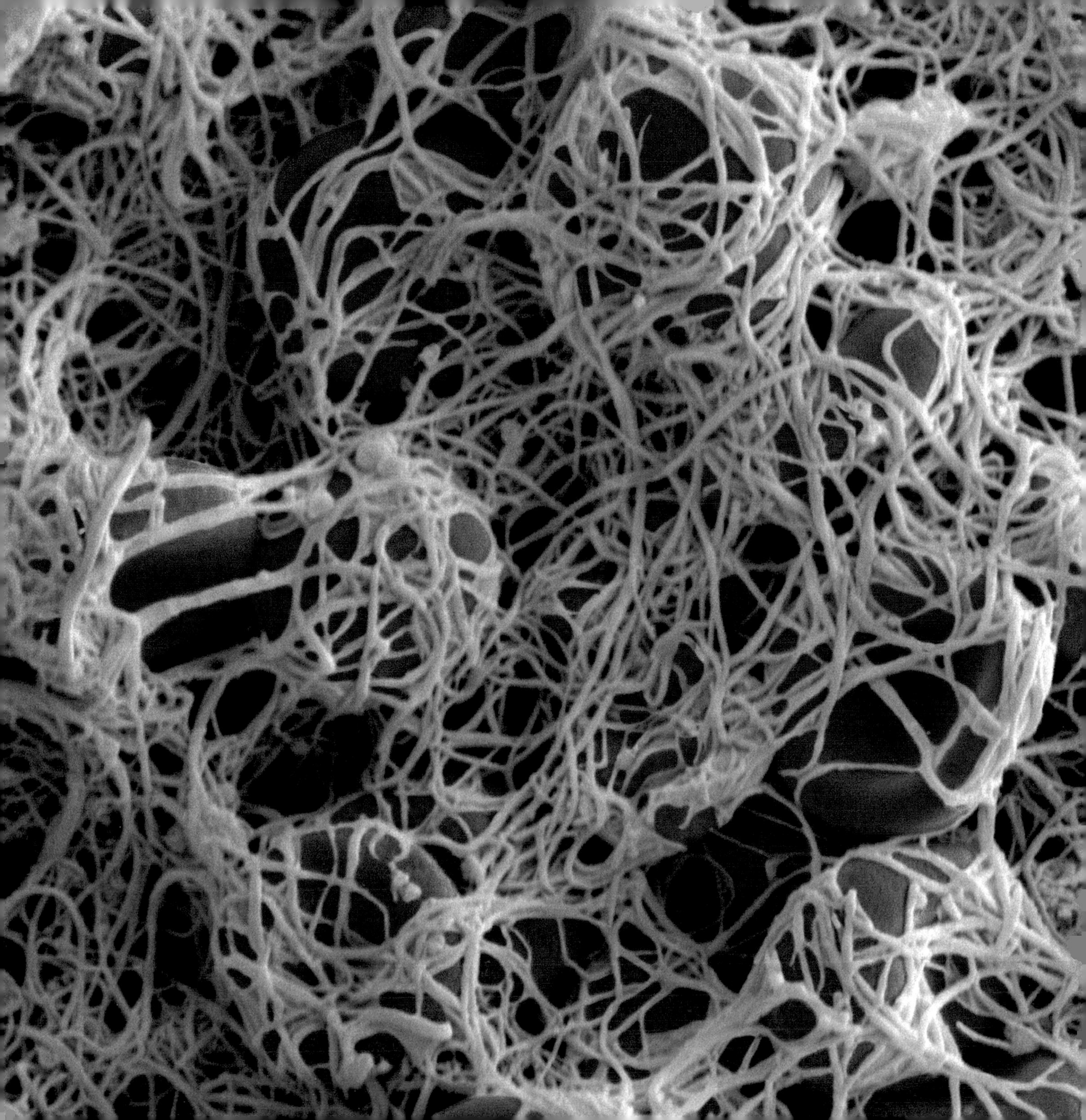

of certain basic functions (diction, mobility). Generally, roughly one fourth of stroke victims die within the following year (see Figure 5), and most of those who survive remain seriously handicapped.

Strokes were first described by Hippocrates 2,400 years ago, who used the term apoplexia (from the Greek for "struck violently"), referring to the death or sudden paralysis of people struck down by them. For a long time, the origin and exact nature of apoplexies were not understood by doctors. It was only in the 17th century that Swiss physician Jacob Wepfer explained, in his *Historiae apoplecticorum* (1658), that apoplexies were connected to bleeding in the brain as well as to a blockage in the vessels irrigating that organ.

The stoppage of blood flow in the brain may be caused by two distinct kinds of stroke, ischemic strokes (also called cerebral infarctions) and hemorrhagic strokes. Ischemic strokes make up 80% of cases, but cerebral hemorrhages (which make up the other 20%) are the most deadly.

Cerebral ischemia occurs following the blockage of a blood vessel in the brain or neck that prevents blood from reaching a region of the brain. As with myocardial infarction, this occlusion is generally the result of the rupture of plaque on the wall of the vessel leading to

Figure 5

Source: www.pdrhealth.com

the formation of a clot (thrombus) that completely blocks blood flow. These clots may also be formed by debris coming from arteries located in another part of the body and transported by the blood to the brain, where they end up blocking a vessel; this is known as a cerebral embolism. A frequent cause of these embolisms is auricular fibrillation, a type of cardiac arrhythmia in which the heart auricles contract rapidly and irregularly. These conditions can cause the blood to form clots that can subsequently affect circulation in the brain.

Cerebral hemorrhages, on the other hand, are caused either by the post-traumatic rupture of an artery in the brain or by damage to blood vessel walls. This damage can be caused by long-term hypertension or be a secondary deterioration related to certain lifestyle habits, such as smoking. This rupture can be catastrophic, as not only does it interrupt the blood supply, essential to maintaining brain functions, it also causes bleeding and accumulation of blood in brain tissue. When the bleeding occurs inside the brain itself (intracerebral hemorrhage), the nearby cells are damaged by the sudden increase in pressure from the excess blood. When the rupture occurs in regions involved in the control of basic functions, like the brain stem, death can occur rapidly. Cerebral hemorrhages can also happen when a vessel is damaged in the space between the brain and the cranium (subarachnoid hemorrhage) following, for example, an aneurysm. The abnormal presence of blood in this area increases intracranial pressure and causes sudden headaches. Almost half of those who suffer this kind of stroke die within 2 weeks of the rupture of the vessel, and a third of survivors need care for the rest of their lives.

Cancer: Cellular mutiny

In his very pessimistic work *Civilization and Its Discontents* (1929), Sigmund Freud states that civilizations are ruled by two opposing kinds of impulses, Eros (the drive for desire and love) and Thanatos (the drive toward destruction and death), with the latter always trying to take control and cause the self-annihilation of society. The validity of this interpretation of the dynamics at work in human societies can be discussed at length. However, in terms of the "cellular civilization" that is the human body, there is no doubt that this is an excellent psychoanalysis! While "cellular Eros" is vital to hold the human body together and keep it working harmoniously, this equilibrium is constantly threatened by the cells' natural tendency to break rank and regain their freedom of action, a "cellular Thanatos" responsible for the many disturbances that cause cancer.

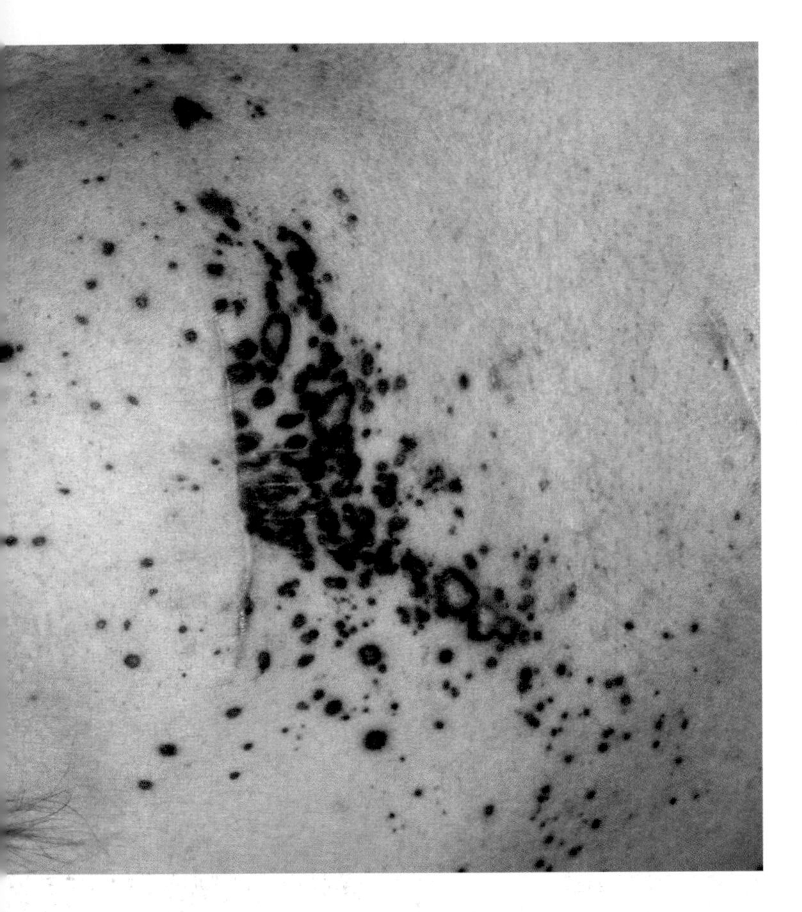

What we call "cancer" is in fact a generic term encompassing more than 200 distinct diseases, all caused by the uncontrolled growth of the body's cells. Having become in recent years the main cause of death in several parts of the world, cancer is a disease that causes fear, as much because of its devastating potential as for the suffering it imposes. If we could eliminate one disease as a cause of our death, most would put cancer at the top of the list.

Dying of cancer is the end result of a long process during which the cells in a given organ undergo changes in their genetic material (mutations) that trigger a radical change in their behavior toward the organism's other cells. Whereas cells are normally specialized to carry out a specific function in the organ where they are found (the function of a skin cell is very different from that of a neuron cell or a pancreatic cell, for example), the accumulation of mutations destroys this identity and returns the cell to a non-specialized state, where its sole preoccupation is the need to reproduce. Cancers are monoclonal in origin, which means they only develop from a cell that has accumulated enough mutations to become invasive and hyperproliferative. This transformation is really a mutiny against the organism: the functioning of an organism as complex as the human body, composed of thousands of billions of cells, requires absolute dedication on the part of all cells to the maintenance of its equilibrium, with each cell having to remain faithful without exception to the role for which it is programmed. This loss of specialization is cancer's ultimate signature. The cell tears up its social contract with the body's other cells and reverts to being neo-embryonic and nonspecialized. Cells that are functionally atypical are

^ Melanoma in a human being.

also morphologically and cytologically atypical, and this atypicality is a major criterion used by pathologists in diagnosing cancer.

Even though everyone knows that cancer is a potentially fatal disease, the ways in which it causes death in those affected are generally poorly understood. Cancer is often viewed as a disease that must be "fought" against, a battle whose end result depends mainly on the energy and determination of the person who is ill. Although this psychological aspect must not be ignored, particularly because it often helps patients to better tolerate treatments or accept the inevitability of their impending death, dying of cancer is certainly not a sign of weakness, as this disease really does have enormous destructive potential. It is not force of will nor the desire to live that will enable the patient to survive, but a group of individual and clinical factors: the patient's state of health before the cancer diagnosis and general metabolic condition, genetic variations that affect the response to chemotherapy drugs, and the ability of the immune system to resist circumstantial and other infections.

Despite the range of organs that can be attacked by cancer and the many disorders caused by this disease, two main mechanisms stand out as responsible for its destructive potential.

Direct effect This is the loss of function in the affected organs. Since cellular specialization

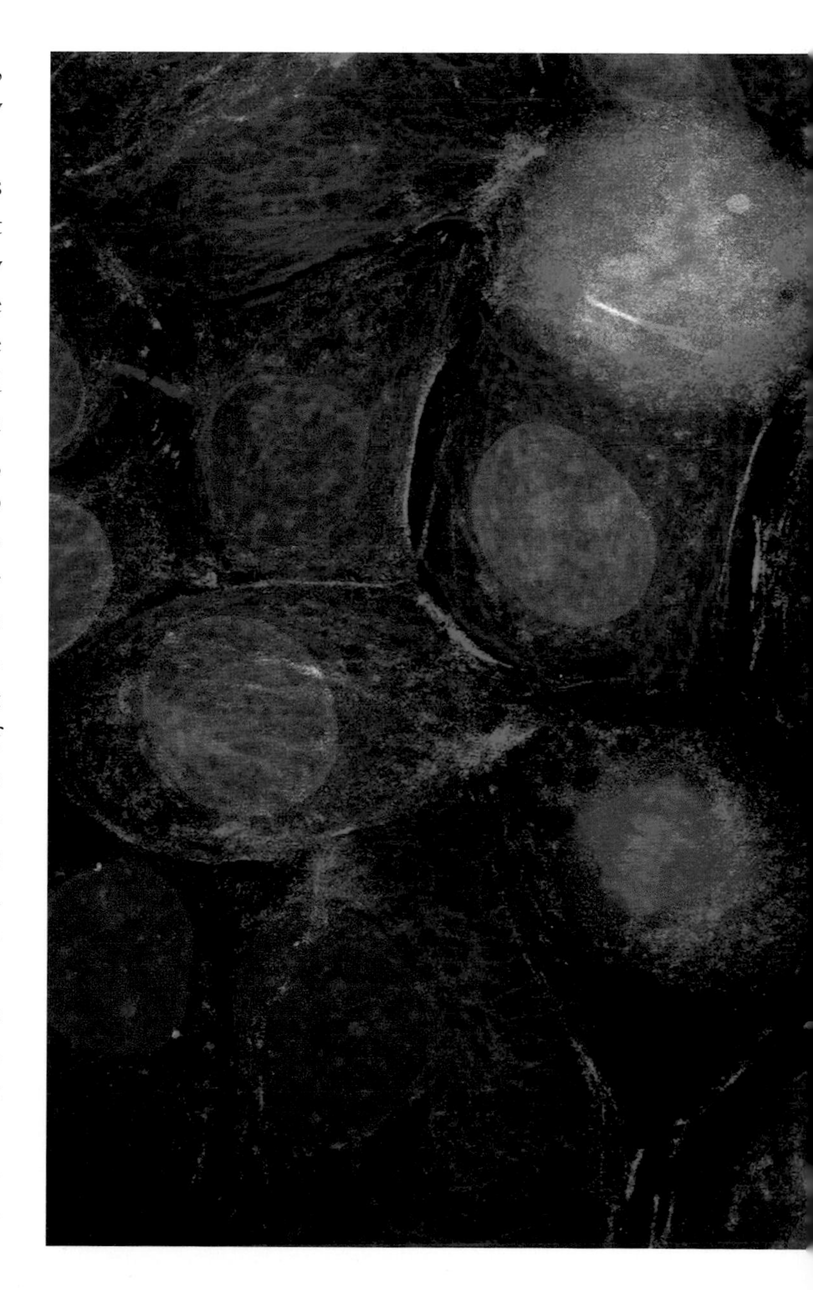

^ Microscopic image of cancer cells in a breast.

煮取二合温服乳癰久不愈者外以膏著
用快刀割瘡孔去腐物敷破敵膏拔之葛根加木附湯兼以
大黄牡丹皮加薏苡仁湯
癩病
一匁或二匁丸茶用

is a prerequisite for a given organ to carry out its specific functions, the loss of this specialization as a result of the dedifferentiation of cancer cells has, in the long term, catastrophic consequences for the functions normally carried out by this organ. For example, if this loss of function occurs in the lungs, the decrease in the number of cells able to capture enough oxygen from the air causes the level of blood oxygenation to drop dramatically, until it reaches a point of no return and the amount of oxygen is insufficient for vital functions to continue. In some cases, the dangers presented by cancer are essentially mechanical in nature. Thus, a cancerous mass in the colon or on the ovaries obstructs the digestive system and prevents essential nutrients in food from being absorbed. In some leukemias, astronomically high levels of white cells in the blood increase its viscosity so much that circulation becomes impossible, and the growth of cancer cells in the brain eventually compresses certain areas governing basic functions until they lose their ability to function, with death as the result.

The body's enormous potential for adaptation ensures that it very often manages to maintain its most important vital functions despite the presence of cancerous cells. This indeed is why a cancer can grow undetected for several years, without causing any particular symptoms. A brain or kidney tumor, for example, can reach a considerable size without significantly affecting the functions carried out by these organs. When it reaches a certain stage, however, the cancerous mass becomes too large to remain hidden, and the disease's first physical symptoms (a palpable mass, bleeding) or metabolic symptoms (loss of appetite, weight loss) begin to appear. Some sudden cancers, such as those in the brain, kidney, liver, ovaries or pancreas, often leave those affected with very little chance of survival, since the cancerous mass has already reached an advanced stage by the time the first clinical symptoms appear.

Indirect effect This is cancer as a systemic disease. While extremely dangerous, the growth of a tumorous mass in the particular organ where the original tumor is located is not the main cause of deaths associated with cancer. In the majority of cases, cancer cells must be present in several locations in the body for the cancer to become powerful enough to take control of vital functions and cause death.

The spread of cancer cells in the organism in the form of metastases is responsible for nearly 90% of deaths due to cancer. This situation stems from cancer's "imperialist" view of the human body—the "founding" tumor usually tries to "colonize" other parts of the organism to compensate for a lack of nutritional resources locally. Such imperialist goals obviously require

< Painting of the first breast cancer surgery carried out under general anesthesia, performed by Seishu Hanaoka in 1804 (artist unknown).

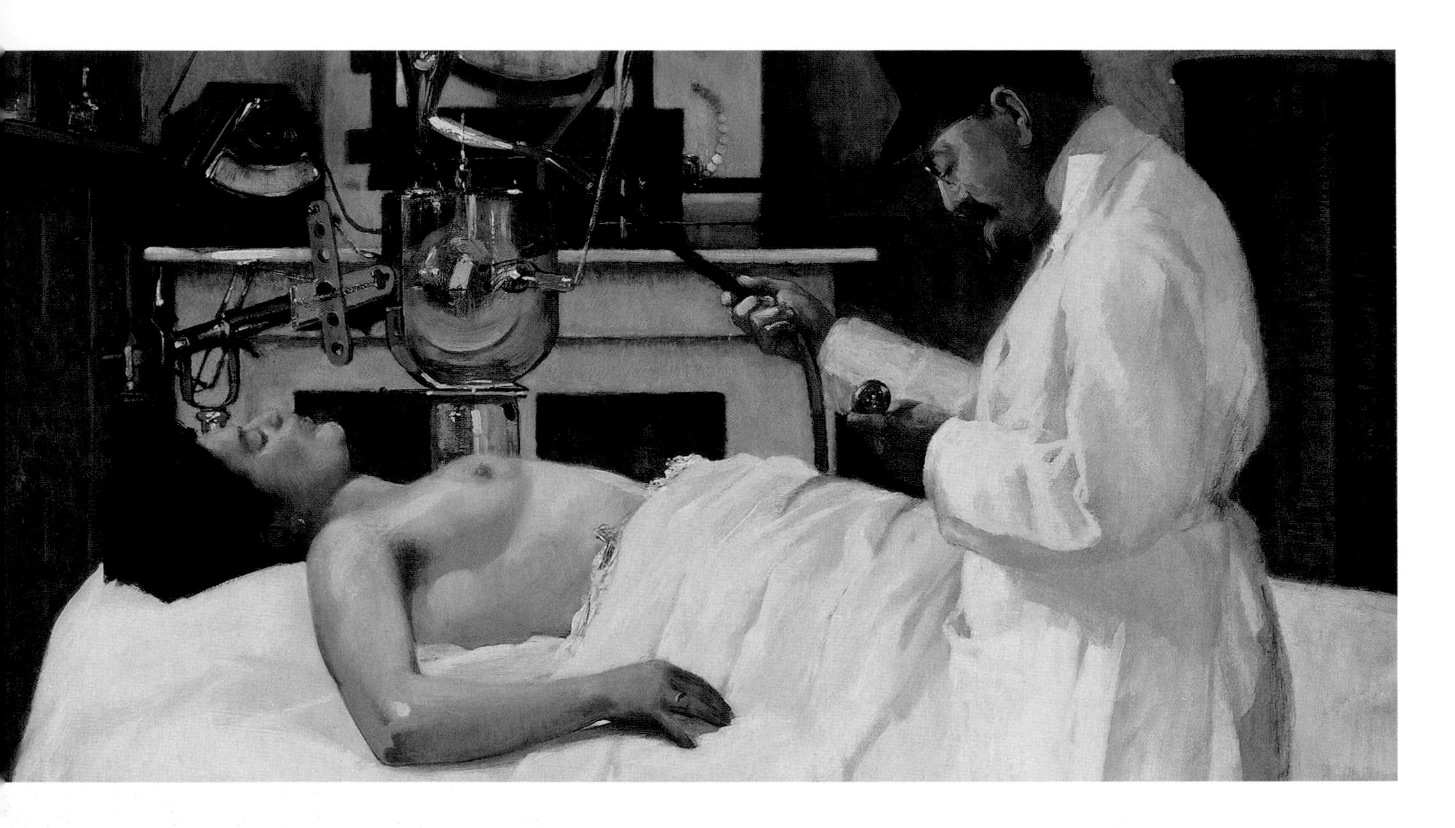

a specialized weapon. Rather like how the British navy made it possible for England to extend its reach across great distances and establish itself in new territories, cancer has perfected a number of weapons for freeing itself from its original site and establishing itself in other parts of the body. One of the key tools used in this expansion is protease production; proteases act just like "molecular scissors," dissolving the tissue around a tumor and thus enabling cancerous cells to go and "explore" the body in search of new sites for the establishment of new colonies.

Even though the organs of choice for the establishment of metastases vary noticeably depending on the location of the primary tumor, the most frequently colonized sites are the lungs, liver, brain and bones. For example, cancers of the colon have a strong tendency to recur in the liver (one patient in four already has metastases at the time of diagnosis), with the cancer cells

^ Georges Chicotot, *The first attempt to treat cancer with X-rays by Dr. Chicotot* (detail)

gradually invading this vital organ until its functioning is affected; this is why loss of liver function is often a cause of death in these patients. The deadly repercussions of metastases are especially obvious when the primary tumor affects a tissue whose function is not essential to life. The best example is indisputably breast cancer, a disease whose fatal potential does not stem from the uncontrolled growth of mammary tissue, which plays no essential role in a woman's biological survival, but rather from the spreading of these cells throughout the organism.

In addition to their ability to form metastases, cancer cells trigger many systemic upsets in the organism, which taken together profoundly disturb the body's functioning and put the patient's life in danger. For example, a high proportion of people suffering from cancer die from the effects of acute renal failure, in which there is an abrupt drop in the kidney's blood filtration rate causing major disruptions in the levels of metabolic wastes (acids, urea) and those of several electrolytes (potassium, calcium, phosphate). Kidney failure may also be a result of chemotherapy treatment for certain cancers (blood cancers in particular). Loss of kidney function is always a disastrous event that leads invariably to death if it is not immediately treated.

Blood clotting disorders constitute additional significant collateral damage associated with the presence of a cancerous mass. First described by French doctor Armand Trousseau (1801–1867)—who observed that people suffering from stomach cancer (himself included) are predisposed to thrombophlebitis—these disorders occur because of the abnormally high level of proteins on the surface of cancer cells. These proteins are able to activate the clotting cascade, resulting in cancer patients having a heightened tendency to develop blood clots in their veins. When these clots reach the heart and lungs, the risk of embolism increases considerably. Far from being a simple curiosity, clotting disorders are in fact a significant cause of death. Nearly one patient in seven dies from complications caused by a pulmonary embolism.

> Mass of cancer cells in the lungs.

The presence of a fast-growing cancerous mass also has profound repercussions for the management of the body's energy reserves. When it reaches a certain size, the tumor enters into direct competition with the organs for the nutrients needed for cell growth. Cancer cells are able to mobilize their host's energy reserves by secreting molecules that speed up the destruction of tissues, particularly fatty tissue and muscles, which often results in a loss of appetite, significant weight loss and loss of muscle mass. This state of thinness combined with extreme weakness, called cachexia, is terrible both for the patients and for those around them, since it leads to a dramatic decline in quality of life; the patient grows ever weaker under the weight of the disease. Once it has reached this stage, the cancer has become a systemic disease affecting the entire body, as if a hideous parasite has taken total control of the organism and hijacked all the functions normally designed to maintain life. That life now hangs by a single thread: the muscles become so weak they struggle to maintain breathing; the metabolism of essential elements is completely disrupted and can barely sustain cell function; and defending against attacks by microorganisms is more and more difficult. In this state of extreme weakness, opportunistic infections can then take hold. Death will occur relatively quickly, depending on individual physiological resistance.

Alzheimer's disease

Although the rate of death from neurodegenerative diseases such as Alzheimer's disease is much lower than the havoc wreaked by cardiovascular accidents and cancer, these diseases are nonetheless terrifying, owing to their ability to attack the most private part of ourselves: our personality. No other natural death is as specific to the human species as that resulting from Alzheimer's disease. The final cessation of the body's vital functions only confirms the death of the body; the death of the person as we knew him or her would have occurred several years previously.

First described in 1906 by German psychiatrist Alois Alzheimer, the disease that has since borne his name begins relatively benignly: the person generally displays short-term memory loss and unusual difficulty in carrying out certain simple routine tasks. As it progresses, however, it reaches the parts of the brain that control language, the emotions and abstract reasoning, and it therefore causes major upheavals in patients' personalities, as they become more and more detached from outside events.

This loss of contact with everyday life is not just a simple dementia associated with an anomaly in the relative levels of neurotransmitters. Alzheimer's disease is truly a degenerative disease in which neurons are completely

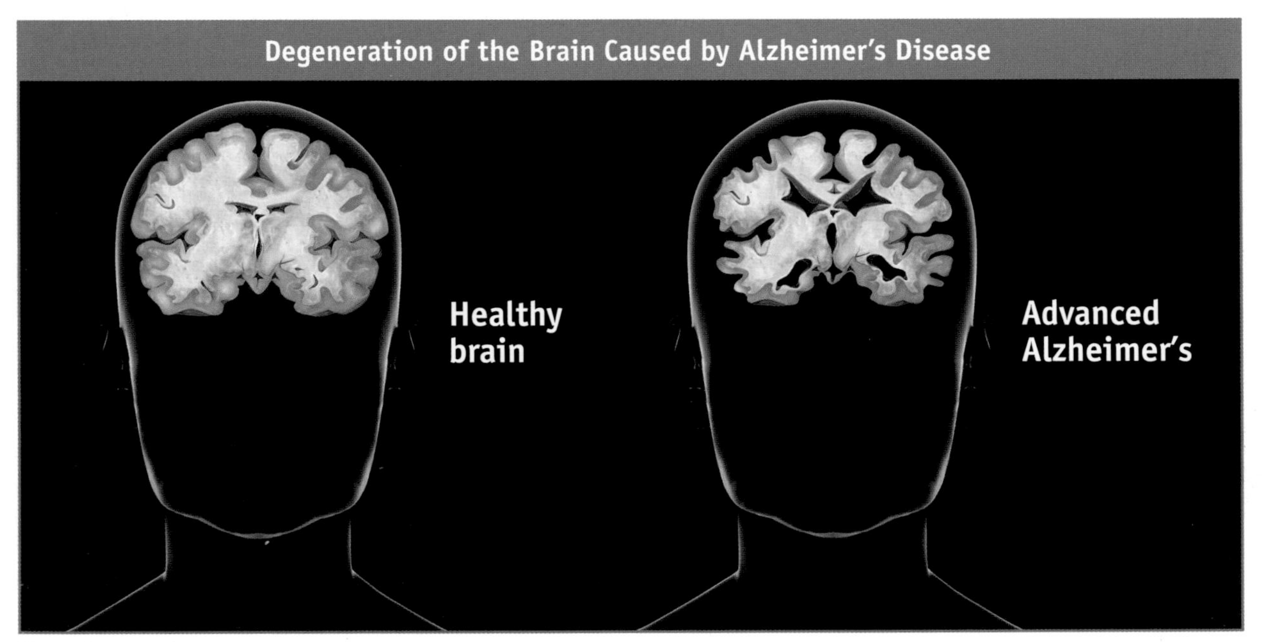

Figure 6

destroyed by accumulated neurofibrillary degeneration and amyloid plaques (or senile plaques). The deposits are created by the aggregation of certain proteins both inside and outside nerve cells and build up gradually; the very wholeness of the brain mass is thus lost (see Figure 6). The presence of these deposits has a disastrous impact on nerve cells, both by considerably increasing their vulnerability to oxidative and inflammatory stress and by directly causing their destruction. In the initial stages of the disease, these aggregates are most commonly found in the regions of the brain involved in memory and emotions (the hippocampus and amygdala), which is why forgetfulness is one of the first symptoms of the disease. Later on, the lesions spread to the associative areas, resulting in the gradual destruction of all acquired knowledge (reasoning, visual recognition, social functioning).

The unrelenting progression of neuron degeneration means that people suffering from this disease lose not only their cognitive functions but also the control of several vital functions governed by the brain. For example, in the advanced stages of Alzheimer's disease, the coordination of breathing and swallowing is greatly

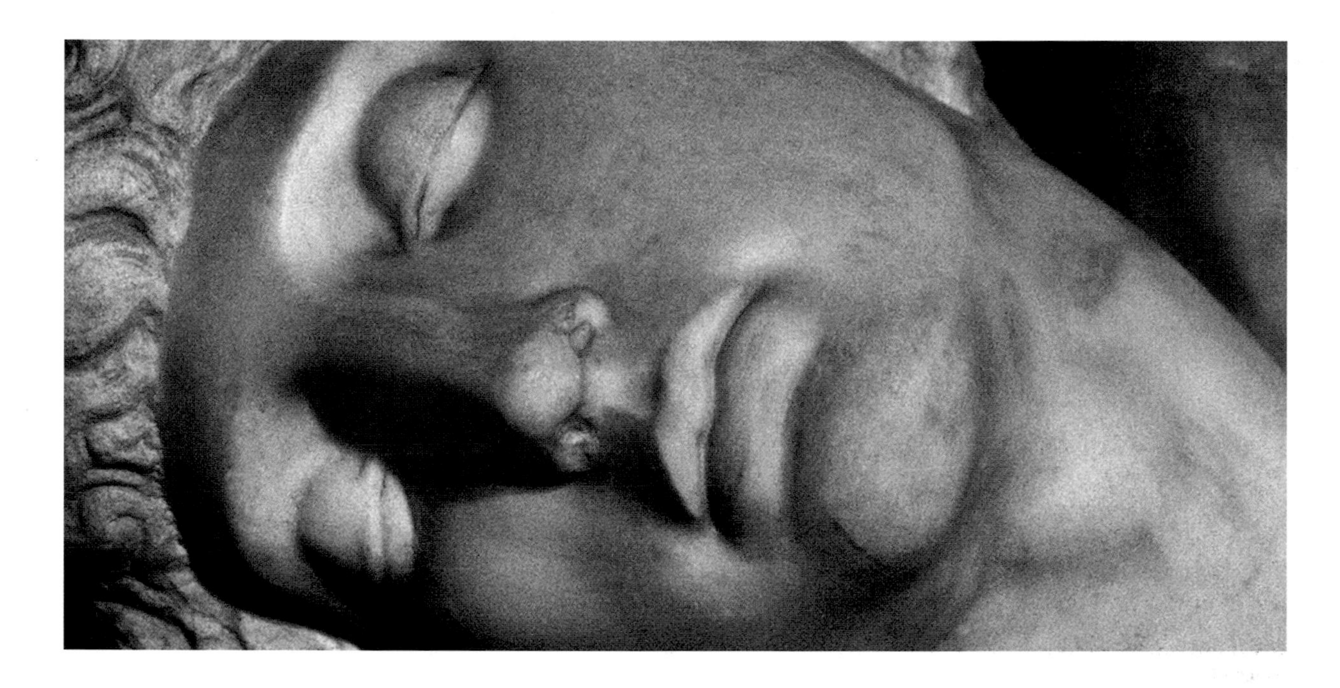

affected, and this often results in foods or liquids being inhaled into the lungs. The abnormal presence of food in the respiratory system provides bacteria with an exceptional source of substances in which they can grow and create an infectious site that attacks the lungs. This aspiration pneumonia is the most frequent cause of death in advanced stages of the disease. But, in addition to this damage to the body, it is without a doubt the death of the personality that makes us fear Alzheimer's so much. The physical deterioration of the brain, which entails the death of a beloved and loving person, of his or her past, life experience and personality, illustrates the importance of brain tissue in our understanding of what it means to be an individual. When it dies, a large part of the person created by this brain activity goes with it.

Euthanasia

In the advanced stages of certain serious diseases, notably cancer, the burden of the disease reaches a point where there is no longer any hope of a cure. In the face of imminent death,

^ Michelangelo, *Dying slave*

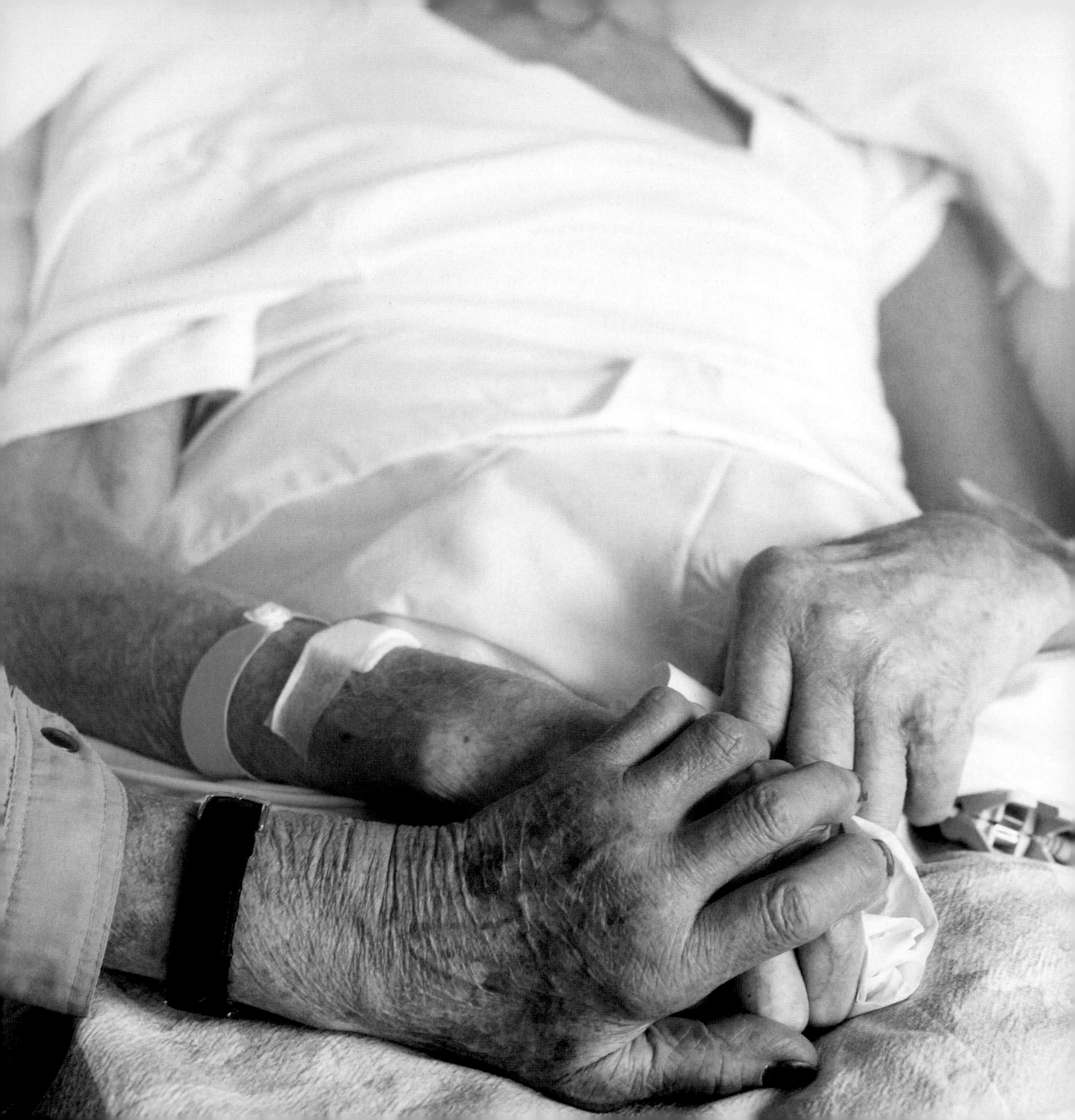

no further attempts are therefore made to save the patient's life; we try instead to alleviate the pain of dying as much as possible, through palliative care (from the Latin *pallium*, "that which protects and comforts"). By alleviating physical suffering with powerful painkillers such as morphine, as well as by offering psychological and spiritual support to patients and their families, this approach tries to improve patients' quality of life as much as possible while allowing the process of life to end naturally, neither speeding up nor postponing death. Some people, however, do not wish to continue to live under these conditions and instead view these last moments as a loss of their autonomy and quality of life and an attack on their dignity. These terminally ill patients demand the right to decide for themselves to end their lives. Others, on the contrary, suggest that life should take its natural course, with everything being done to eliminate as much suffering as we can—in other words, by offering palliative care to the fullest extent possible. The moral, ethical and legal questions raised by the desire of people who are ill to control the final moments of their lives are currently the best illustration of the difficulties inherent in this problem.

Euthanasia, from the Greek *euthanatos*, meaning "good death," may be defined as an act designed to cause the death of an incurably ill person so as to cut short his or her suffering and death throes. It was practiced in ancient Greece and Rome, where it was common for people who were suffering dreadfully to commit suicide, often with the help of poisons provided by their physician. Although this practice was accepted by philosophers such as Socrates, Plato and Seneca, Hippocrates (the father of modern medicine) was firmly opposed to it. His view gradually gained the upper hand in the Western medical tradition, and it is still clearly stated

> Portrait of Hippocrates (c. 460–377 BCE), the Greek physician considered to be the "father of medicine."

today in the oath that bears his name: "I will never deliberately cause death." This opposition grew stronger with the parallel development of a number of religious movements that consider life to be the concrete result of an act of God, a gift that humans cannot just cast aside as they see fit.

The harsh reality of the dreadful suffering endured by patients in the terminal phase has nonetheless always aroused human compassion and has led many to call for a peaceful death to put an end to this suffering.

The complex nature of the current debate over euthanasia is therefore rooted in the influence of all these currents of thought combined, a true moral dilemma caused by the collision of three of humankind's most noble characteristics: the desire to care for those who are ill, the awareness of the unique nature of existence and compassion.

Currently, only the Netherlands, Belgium and Luxembourg authorize active euthanasia in the case of incurable disease, according to a procedure that is in all cases tightly regulated. When a patient requests it, the physician must be sure that the suffering is intolerable, the disease is incurable and the desire to end one's life is unequivocal. He also has to consult with at least one other independent doctor. As patients are sometimes unable to express their wishes, the law allows anyone to state in writing his or her wish to be euthanized in the event he or she becomes ill with an incurable disease, goes into a coma and is in a state deemed to be irreversible.

If the request for euthanasia meets legal requirements, the patient's death is generally brought about by intravenously administering a sedative called sodium thiopental. Once the patient is in a deep coma, a powerful muscle relaxant, pancuronium bromide, is administered to stop breathing and thus cause death. In the Netherlands, between 6% and 10% of cancer patients in the last stage of their illness choose to die in this way.

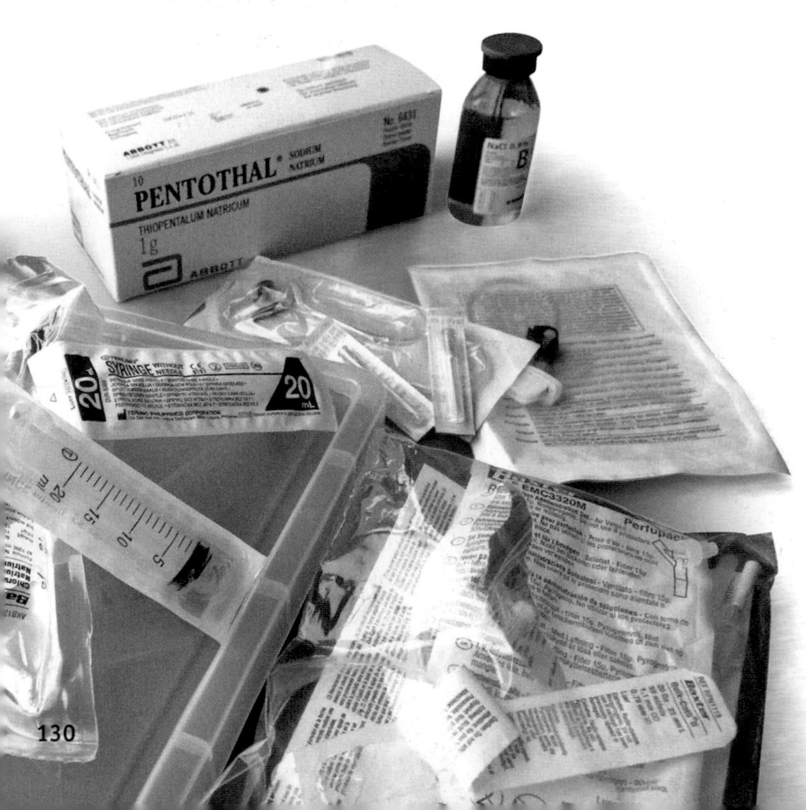

< Components of a euthanasia kit available in April 2005 in Belgium for physicians wishing to practice euthanasia at home.

Euthanasia must also be distinguished from assisted suicide. In the case of the latter, even if the physician prescribes the lethal substance and advises the sick person on the procedure to follow, it is the patient who administers the fatal dose, without any medical assistance.

Assisted suicide is legal in Switzerland and in Oregon, Washington and Montana, even though the laws of all these states expressly prohibit all forms of active euthanasia. As is the case for euthanasia as practiced in Belgium and the Netherlands, the procedure to be used in assisted suicide is tightly regulated. In Oregon, for example, patients wanting to obtain a lethal substance must be at least 18 years of age, be suffering from an incurable disease with fewer than 6 months to live and be able to clearly and precisely express to a physician their decision to end their life. The doctor may then prescribe a lethal dose of sodium pentobarbital or sodium secobarbital, two barbiturates that are powerful central nervous system sedatives. A few minutes after ingesting either one of these substances, the patient sinks into a deep coma, and the subsequent paralysis of respiratory functions usually causes death in under 30 minutes. Since 1998, approximately 40 people each year have resorted to assisted suicide in Oregon.

An ethical problem

It is difficult to arrive at a consensus on euthanasia or assisted suicide, as not everyone holds the same view of free will with respect to the last moments of our lives. Although the debate over

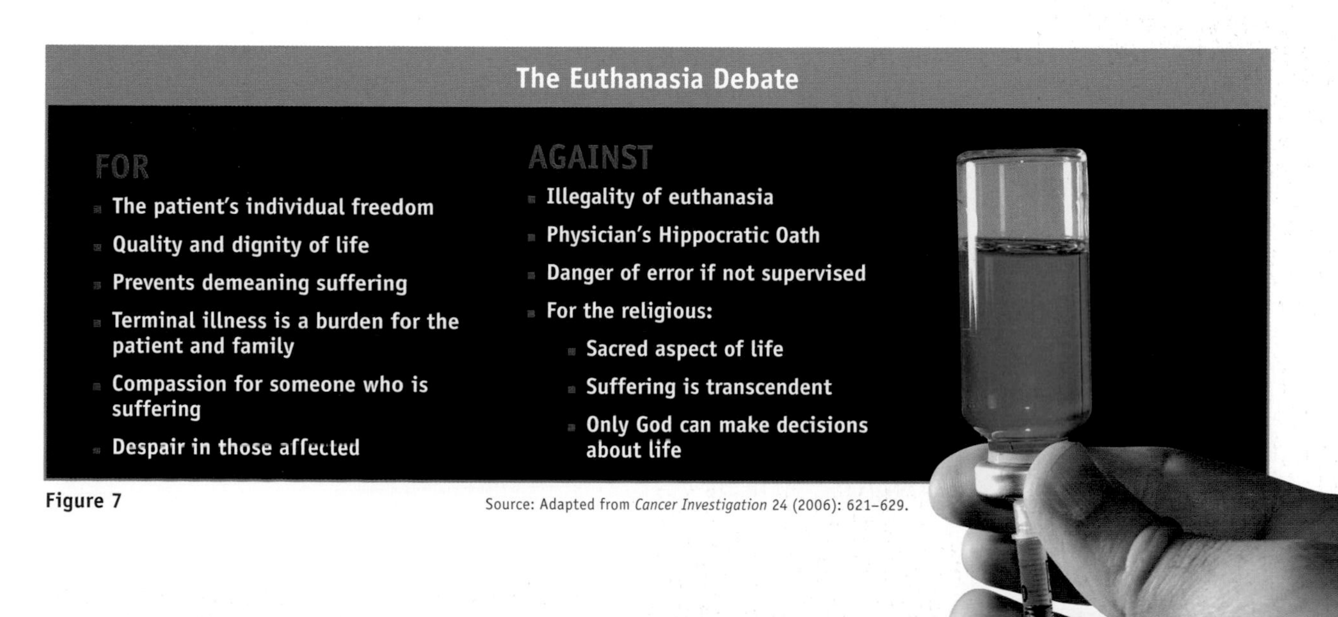

Figure 7

Source: Adapted from *Cancer Investigation* 24 (2006): 621–629.

euthanasia is a very old one, it is interesting to note that the arguments used to defend or condemn this practice have remained essentially the same for centuries (see Figure 7).

For its advocates, the possibility of choosing how we live our last moments is a fundamental individual right: anyone wishing to end his or her life because he or she does not accept the decline in quality of life and the burden placed on family members should be able to request active (euthanasia) or passive (assisted suicide) assistance from a doctor. Conversely, opponents of euthanasia consider it to be murder, whatever the circumstances, and therefore a violation of our fundamental right to life. In some religions, God is the absolute master of His creatures, and human beings cannot oppose His will. In this case, the suffering endured in the final moments of life is not viewed from a strictly negative perspective, but rather as a major test that gives us the chance to reflect on the meaning of life and make peace with ourselves.

In societies where religion is omnipresent and often inseparable from the power of the state, the issue of euthanasia and assisted suicide is more easily resolved, since the sacred nature of life usually takes precedence over other considerations. However, the situation is much more complex in secular societies. Although religion is in the private sphere and separate from the state, it has often played a predominant role in the development of laws and still has a decisive influence on moral and legal values. This is why, even though we grant citizens freedom of action, whatever his or her ethnic heritage, nationality, sexual orientation or religion, this freedom does not include, in most countries, the right to resort to medical assistance to end one's life in the event of incurable illness. In Canada, for example, euthanasia runs counter to Article 7 of the Canadian Charter of Rights and Freedoms, since the law states that the right to life takes precedence over any medical procedure that aims to shorten life, even when there are no further therapeutic means for saving the patient.

Euthanasia and assisted suicide are two serious ethical problems that are important to consider. How do we chart the new pathways blazed by an aging population and by legitimate questioning about the meaning or non-meaning of suffering? How can we respect individual freedom, essential to our modern societies, while protecting the most helpless?

> Bruce Turner, *The Dying Man*

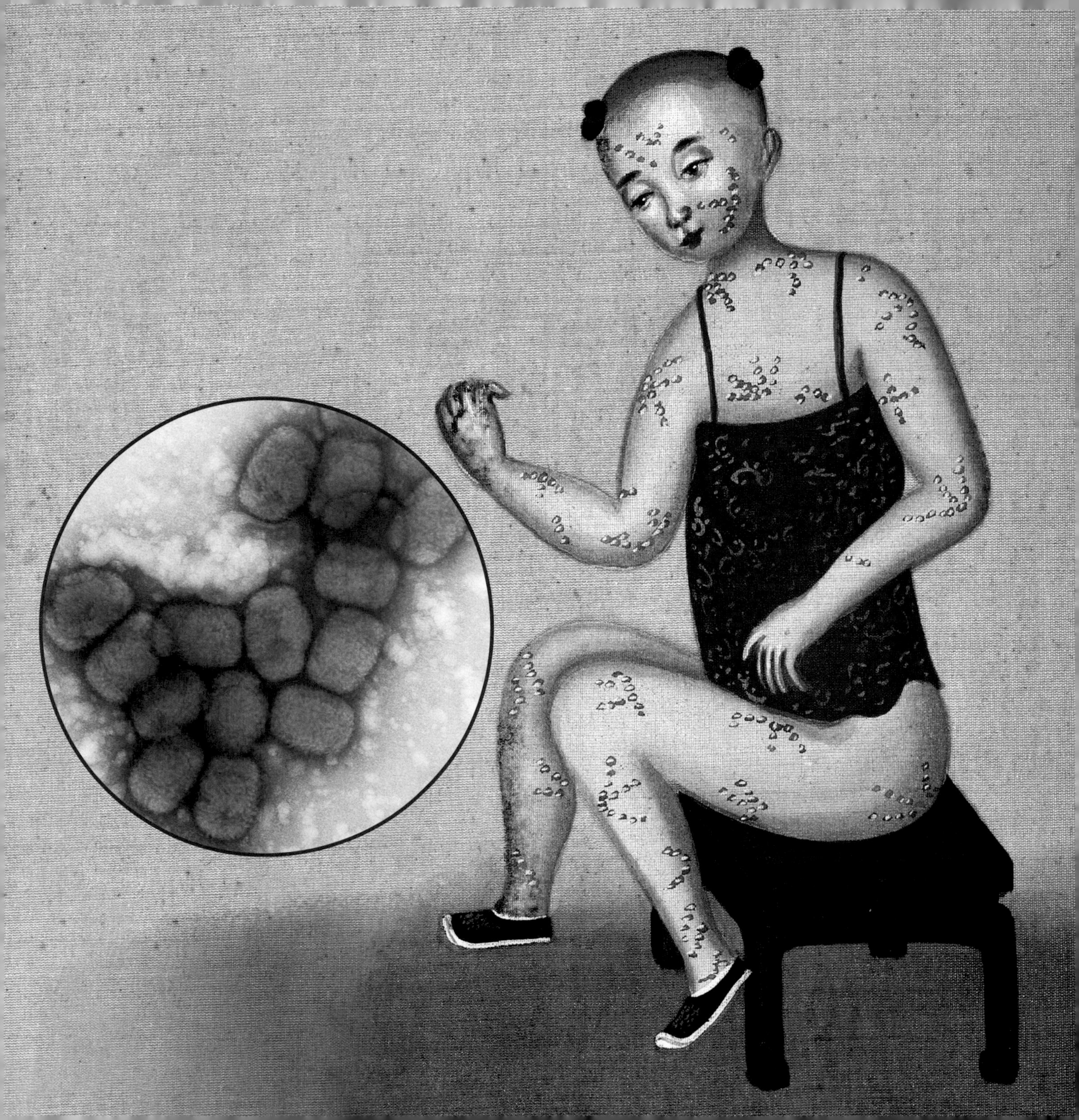

Chapter 6

Dying from Infections

> Gentlemen, it is the microbes that will have the last word.
>
> Louis Pasteur (1822–1895)

In one of the most famous stories in Greek mythology, Pandora, devoured by curiosity, defies Zeus's prohibition and opens a sealed box, received as a gift upon her marriage to Epimetheus. Great was her misfortune, for immediately there escaped all the evils of humanity—Old Age, Disease, War, Famine, Poverty, Madness, Vice, Deceit and Passion, and these scourges spread all over the world, bringing suffering to mortals. Terrified, Pandora hurriedly closed the lid, but unfortunately it was too late: only Hope remained at the bottom of the box.

As a poetic metaphor that tries to explain daily trials as the consequence of the weakness of humankind and the power of supernatural beings, this legend attests to the distress human beings feel when confronted with tragic events.

Of all the scourges that escaped from Pandora's mysterious box, diseases, and more especially infectious diseases, are without a doubt the greatest ordeals that human civilizations have had to face. Plague, smallpox, syphilis, tuberculosis, measles, malaria, cholera, influenza, AIDS and the several hundred other diseases caused by bacteria, viruses and parasites have, throughout history, left in their wake suffering, grief and terror, prematurely cutting short innumerable lives and decimating populations—even entire civilizations.

The almost instinctive fear that infectious diseases arouse in us, even today, is rooted in all these calamities, based on knowledge going back

< Child ill with smallpox, artist unknown; (inset) Electron microscope image of the smallpox virus

thousands of years and passed down from generation to generation, each one despairing at its powerlessness to prevent the mass deaths of fellow human beings. These diseases were long considered to be punishment from an all-powerful God who was unhappy with the actions of His sinful and disobedient people. This divine intervention was a way of attributing meaning to completely incomprehensible phenomena. Indeed, how is it that a healthy and robust person can be so quickly laid low by fever and tormented by hideous skin lesions or an excruciating internal pain that empties out his insides and then die suddenly a few days later?

This distress is understandable since, even today, when we have detailed knowledge about the microbial agents responsible for such swift deaths, the sudden appearance of a contagious disease is still the medical event most likely to immediately capture the public imagination. In the first decade of the 20th century alone, SARS, H5N1 avian flu and the H1N1 pandemic have attracted unusual public and media attention, an interest that turned out to be exaggerated given the relatively low number of deaths caused by these diseases.

But, aside from these exceptional headline-grabbing examples (publicized largely because these diseases threatened the inhabitants of industrialized countries), fear of infectious dis-

^ Greek sculpture of Pandora and her box

eases is fully justified, as even today they pose a real danger to people in most parts of the world. According to World Health Organization statistics, diseases caused by viruses, bacteria or parasites kill more than 14 million people worldwide each year, representing nearly 20% of all deaths. In addition to AIDS, tuberculosis and malaria, which alone are responsible for more than half of these deaths, this category includes many diseases in which diarrhea is a symptom as well as a number of serious tropical parasitic diseases (malaria, trypanosomiases, etc.) and diseases like influenza, which kills nearly 500,000 people a year. Nor should it be forgotten that almost 15% of cancers, some of which are virtually incurable, are bacterial or viral in origin and are a major cause of death in some parts of the world. Pasteur's pessimistic statement quoted at the beginning of this chapter is perhaps not that far from reality: microbes really are the human species' biggest predators, real killers with huge destructive potential.

Conquered by a virus

The infectious diseases brought by the Spanish conquistadors played a major role in the downfall of the Inca and Aztec empires. Indeed, despite their military superiority, it is very unlikely that the Spanish conquistadors could have succeeded in conquering the Aztecs so quickly if they hadn't brought with them a formidable weapon: smallpox. Whereas Europeans had acquired a degree of immunity to this disease over the centuries, the indigenous peoples of the Americas had never come into contact with the virus causing this infection (*Variola major*). No fewer than 19 epidemics ravaged the population in barely a century. For example, whereas there were 1.2 million inhabitants in the Mexico City valley alone in the 16th century, there remained just 70,000 in 1650. This slaughter evidently inspired English general Sir Jeffrey Amherst in his conquest of Fort Carillon. To prevent the Native Americans from helping the French defend the fort, Amherst gave them blankets infected with smallpox. This caused an epidemic that decimated the native population and made it possible for Amherst to take the fort.

> Child with the smallpox virus

The microbe empire

The term microbe, which literally means "little life," includes all organisms invisible to the naked eye, whether they are bacterial, viral or parasitic in origin. Although microbes are among the first species to have emerged from the primal soup more than 3.5 billion years ago, it was only with the invention of the microscope in the 17th century that we were first able to see them. While until then they had "traveled incognito," it is now known that microbes are the planet's most plentiful organisms by far, both on account of their numbers and their diversity, given that there are several thousand distinct species of them (see Figure 1).

While the vast majority of microbes are completely harmless to us, some are very dangerous, as the many deaths caused by infectious diseases throughout history and even today remind us.

Since infectious diseases must by definition be transmitted from one person to another, they appeared quite recently in human history, coinciding in the vast majority of cases with our adoption of a sedentary way of life in settled communities with high population density. Analyses of the skeletons of nomadic hunter-gatherers from the Paleolithic period (e.g., cavemen) indicate that they were not generally susceptible to infectious diseases. The marked increase in population numbers that accompanied the Neolithic revolution coincided, however, with a veritable explosion of diseases as serious as malaria, tuberculosis, poliomyelitis, measles, German measles, smallpox, influenza and plague. Population concentration in communities of ever greater density encouraged the spread of these diseases and opened the door to the disease carried by rodents, who were attracted to food supplies. And this does not take into account the fact that sanitary conditions have not always been what might be called optimal, especially in the Middle Ages, which created an ideal environment for the spread of many deadly epidemics (see box on p. 142).

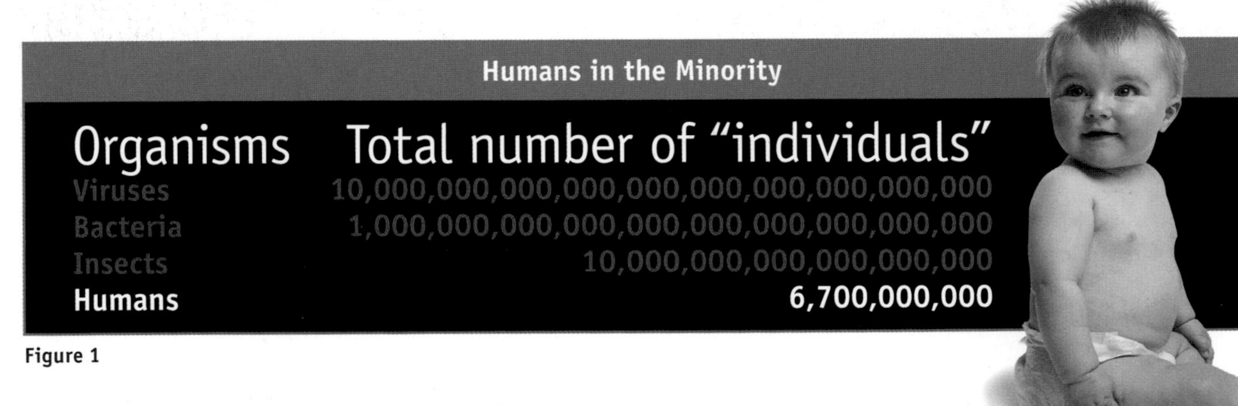

Humans in the Minority

Organisms	Total number of "individuals"
Viruses	10,000,000,000,000,000,000,000,000,000,000
Bacteria	1,000,000,000,000,000,000,000,000,000,000
Insects	10,000,000,000,000,000,000
Humans	6,700,000,000

Figure 1

^ Louis Pasteur, the French scientist whose research on vaccines led to a major medical breakthrough.

^ In 1928, Sir Alexander Fleming discovered penicillin, an antibiotic that revolutionalized the treatment of bacterial infections.

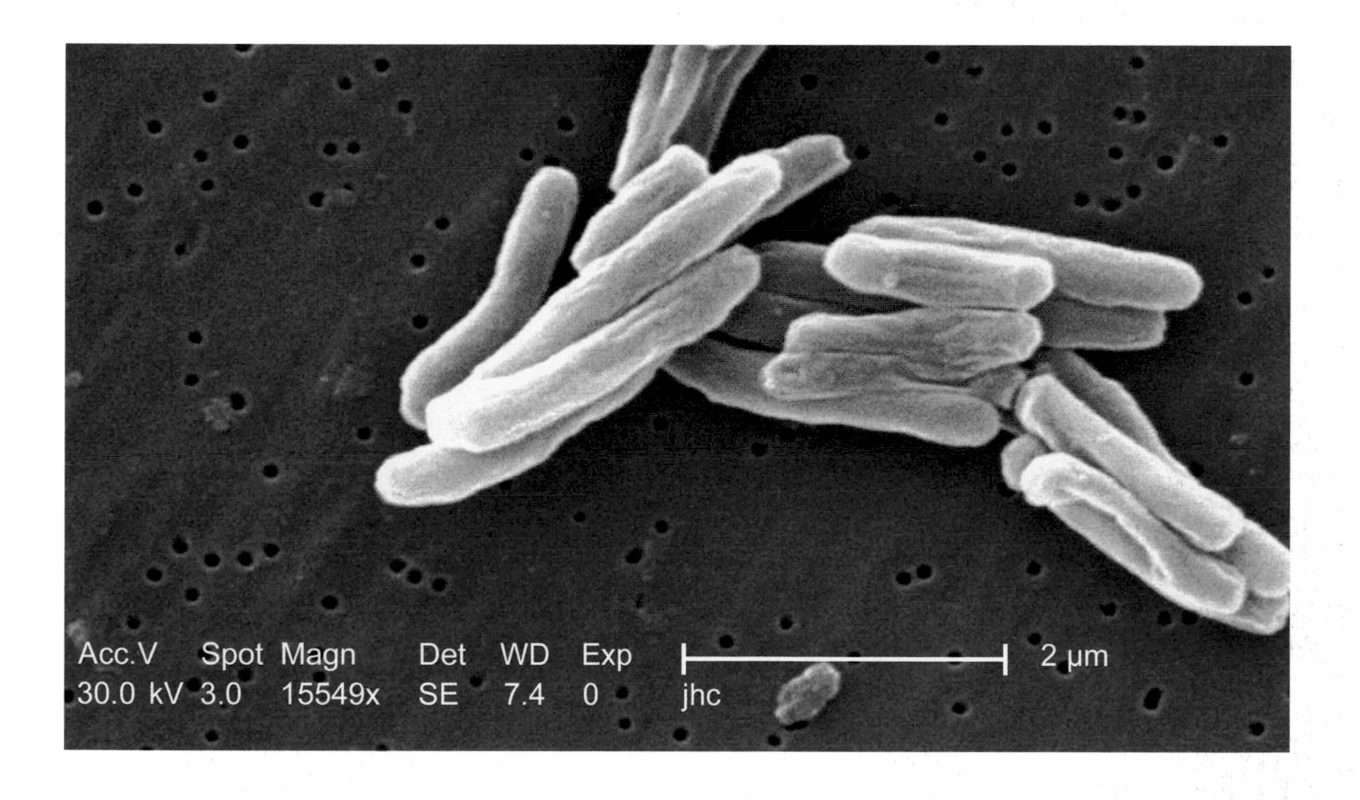

Extreme bacteria

A single drop of seawater contains nearly 10 million viral particles, and ⅓ ounce (8.5 g) of soil from your garden may contain 8.5 billion bacteria. The world of microbes is truly a world unto itself, an extremely complex universe that seems to have managed to use the some 3.5 billion years that have elapsed since their creation to diversify and colonize every corner of the earth. As proof, archaeobacteria have been discovered living in extreme conditions: *Sulfolobus acidocaldarius* in acidic hot springs (185°F/ 85°C); *Halobacterium salinarum* in extremely salty bodies of water (this microorganism gives the Dead Sea its pinkish-red color); and even bacteria that produce methane, active almost 2 miles (3 km) underwater, in samples of ice from Greenland. Fortunately, most of these microbes are completely harmless to humans!

< Robert Kòch discovered the tuberculosis and cholera bacilli; he was awarded the Nobel Prize for medicine in 1905.
< Electron microscope image of the tuberculosis bacillus (*Mycobacterium tuberculosis*).

Unhealthy living conditions and diseases

The word "hygiene" comes from *Hygia*, the Greek goddess of health and cleanliness. The daughter of Asclepios, the god of medicine, she was considered to be a powerful goddess, playing an essential role in maintaining health—a totally appropriate cult given the critical role of hygiene in preventing infectious diseases! Whereas the civilizations of antiquity (Egypt, Greece, China and the Roman Empire) were generally concerned about hygiene and the cleanliness of their cities, the Middle Ages was the golden age of unhealthy living conditions. Although people were relatively clean thanks to the many public baths, "waste management" was absolutely dreadful. In cities, excrement was thrown right from the windows into the streets, where it accumulated along with that of the animals that wandered around freely. And although people took care to shout "Watch the water" or "Watch out below," managing to stay clean while walking in the streets was nothing short of a miracle...

Although a foul-smelling odor was typical of all the streets in Europe, the appalling situation in Paris remains one of the most frequently written about. Since there were no sewers, the muddy streets were full of filth, especially around the slaughterhouses and butcher shops, where the excrement, blood and internal organs of slaughtered animals lay all over the ground and in the gutters. And in spite of several royal edicts imposing measures designed to clean up this unhealthy environment, it was not until after the enormous urban reconstruction projects supervised by Baron Haussman in the second half of the 19th century that these efforts were successful.

Poor hygiene promotes contagion, not only because the presence of organic wastes encourages the proliferation and transmission of disease-causing microbes (cholera bacillus, for example), but also because it provides abundant food for certain disease carriers like rats (which carry the plague). More than any medical discovery, basic improvements in sanitary conditions are the main reason for the decline in infectious diseases in the 20th century.

Murders under the magnifying glass

How can microbes that are invisible to the naked eye strike down a human being in as little as a few hours? The diseases caused by various microbes are too numerous for us to be able to describe in detail all the ways in which these microorganisms can cause death. Nonetheless, some of them strike the imagination more than others owing to their devastating power, which has made them the greatest killers in human history. Historically, plague and cholera are without a doubt the best examples of the ravages infectious diseases can cause (see box on pp. 143–147); today the biggest threats to the inhabitants of industrialized countries come from the influenza and AIDS viruses.

The plague

A bacterium, rodents, fleas and men: much of our fear of infectious diseases stems from the dreadful trauma caused by the plague epidemics that swept across Europe and Asia in the last two millennia. The epidemic we are most familiar with is that of 1347–1351, the infamous Black Death, which, began in Caffa, a city located on the Crimean peninsula (the Black Sea) and affected almost all of Europe, leaving in its wake at least 25 million dead. The many accounts of this period speak of the desolation caused by the disease:

> What with the fury of the pestilence, the panic of those whom it spared, and their consequent neglect or desertion of not a few of the stricken in their need... between March and the ensuing July upwards of a hundred thousand human beings lost their lives within the walls of the city of Florence.... How many brave men, how many fair ladies, how many gallant youths, whom any physician, were he Galen, Hippocrates, or Sculapius himself, would have pronounced in the soundest of health, broke fast with their kinsfolk, comrades and friends in the morning, and when evening came, supped with their forefathers in the other world! (Boccaccio. *The Decameron*, 1348–1353, Trans. by J.M. Rigg, 1903.)

It was pure evil, brutal and pitiless, which left physicians completely at a loss, sometimes having to resort to surprising means to carry out their duties (see "The plague fighter," p. 146). This "Great Death" shook European civilization and, in a number of ways, had a determining influence on the very course of history.

The plague is caused by a bacterium (*Yersinia pestis*) that in its natural state infects rodents, using fleas as carriers. The mechanism is very ingenious. By multiplying in the flea's digestive system, the bacterium gradually blocks its esophagus, preventing the flea from getting the food it needs. Starving, the flea bites all the rodents it can find but without any success, since, each time, the blood it sucks out is blocked by the "plug" of bacteria located at the entrance to its stomach. It therefore regurgitates into the wound the blood it cannot ingest, at the same time transmitting the bacterium to its host. Getting hungrier and hungrier, the flea proceeds in this way from rodent to rodent, always unable to get enough food but nonetheless spreading the bacterium to many new hosts.

Human beings are only an accidental intermediary in the plague's chain of transmission, and it is only when they live in proximity with rodents, as is the case when sanitary conditions are poor, that they are involved in the bacterium's reproductive cycle. When a flea carrying the disease bites a person and the bacterium enters the person's bloodstream, a group of proteins on the surface of the bacterium neutralize the immune response, and the bacterium is thus able to migrate to the lymph nodes. There it multiplies at dizzying speed and causes very painful swellings in the nodes (buboes) that can grow as large as a small apple. Bacteria in the bubo then travel in the bloodstream to all the organs, sounding an "alarm" to the immune system's inflammation-fighting "squadron," easily recognized by a general feeling of being unwell and a high fever (104°F/40°C). Very quickly, the bacteria in the blood cause clots to form in the small blood vessels, which impedes the circulation of blood in the organs as well as cardiovascular collapse, plunging the patient's blood pressure. Most often (in roughly 70% of people affected), death occurs 3 to 5 days after the appearance of the first symptoms as a

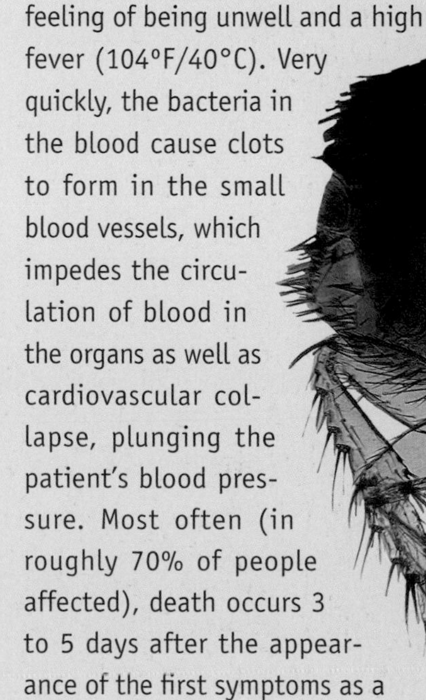

< *Le Triomphe de la mort* (The triumph of death), artist unkown

result of infectious metastases in the liver, spleen, brain and lungs. As Procopius of Caesarea noted during the Justinian plague that struck Constantinople in 543, some of those infected with plague "died vomiting blood." This pulmonary plague is very dangerous, since it is transmitted from person to person, without requiring fleas as intermediaries. After a brief incubation phase (from a few hours to 2 days), the disease begins suddenly with the appearance of hemorrhagic necrotizing pneumonia lesions accompanied by fever (104°F/40°C), coughing (containing bacteria), hemoptysis and breathing difficulties, as well as a profound change in general health. The disease leads quickly to enlargement of the pulmonic areas, neurological disturbances (mental confusion or prostration), subcutaneous hemorrhaging, cardiovascular collapse and death. While a third of those who fall ill with bubonic plague succeed in defeating the disease, pulmonary plague is usually fatal.

The plague fighter

It is to Charles de Lorme (1584–1678), chief physician to Louis XIII, that we owe the creation of the plague fighter's curious costume. The mask, usually white, was composed of a 6-inch (16 cm) long nose shaped like a beak and stuffed with herbs, spices and perfumes, designed to "purify" the surrounding air and (especially!) to block out the dreadful odor given off by the sick and the dead. Under his overcoat made of leather coated in wax, he typically wore leather breeches with boots attached and a plain leather shirt tucked into the breeches. Spectacles for the eyes and a leather hat completed the outfit, giving him a frankly surrealist air. The sight of these *dottore della peste* wandering among the dead and dying victims of the plague and using their long canes to poke the wounds of the sick who were crying out in pain must have added a macabre dimension to the already apocalyptic atmosphere created by the disease.

> Costume of a plague fighter

Cholera

While not as deadly as plague, cholera remains the disease that can kill a human being the fastest: in the most serious cases, it can take the life of a healthy person in barely a few hours. Described for the first time under the name *bisuchika* ("deadly disease of the intestines") in the *Sucruta-samhitâ*, an ancient Indian medical tract, cholera has long been endemic in that part of the world. The disease's infectious agent is the *Vibrio cholera* bacterium, a bacillus that grows especially well in aquatic environments contaminated by garbage left by local populations. This bacterium has the particular ability to produce extremely virulent toxins that pass through the intestinal wall and inactivate key proteins involved in sodium absorption. The consequence of this inactivation is absolutely horrendous, resulting in diarrhea causing large loss of fluids, as much as 2½ to 3 gallons (10–12 L) per day and sometimes even more. This massive loss of fluid causes hypovolemic shock, a state in which the volume of fluid is insufficient to sustain blood circulation. In the advanced stages of dehydration, the person who is sick becomes very pale, with blue-tinged or even blackish hands and feet. This is called cyanosis and results from lack of oxygen in the blood. In fact, the French expression *avoir une peur bleue*—literally to "be blue with fear," or scared to death—originated during the cholera epidemic that struck France in 1832. Cholera is a terrifying disease, the horror of which may continue even after its victims die: in some cases, extreme dehydration of the body causes muscular tension that can make the arms and legs contract and twitch. This characteristic is at the root of folktales according to which it was sometimes the living who were buried, not the dead.

The virus that came in from the cold

The word "influenza" likely comes from *influenza di freddo* (the influence of cold), an Italian expression used in the 18th century to underline the high incidence of this infection in cold weather. We now know that winter is indeed an auspicious time for flu, as the virus that causes it is more easily transmitted when the temperature is low and the humidity level is down, two characteristics of winter weather conditions. In contrast to the cold virus (rhinovirus), which spreads mainly through contact between the hands and the eyes, nose and mouth, the sprays from coughing or sneezing are influenza's main path of transmission. A single cough may release nearly 100,000 suspended particles, and this number can be as high as 2 million for a sneeze. When you consider that it takes just 10 particles to infect a host, the importance of coughing or sneezing into the crook of our elbow when we have the flu is easy to understand!

The influenza family includes three different types of virus, called A, B and C, with type A being the most dangerous by far. Although a number of variants of this virus exist in nature (in pigs, horses and dogs, as well as in several other warm-blooded vertebrate species), birds are the main natural reservoir for the virus, which has been identified in no fewer than 90 different bird species. As in the majority of infectious diseases, it is therefore very likely that transmission of

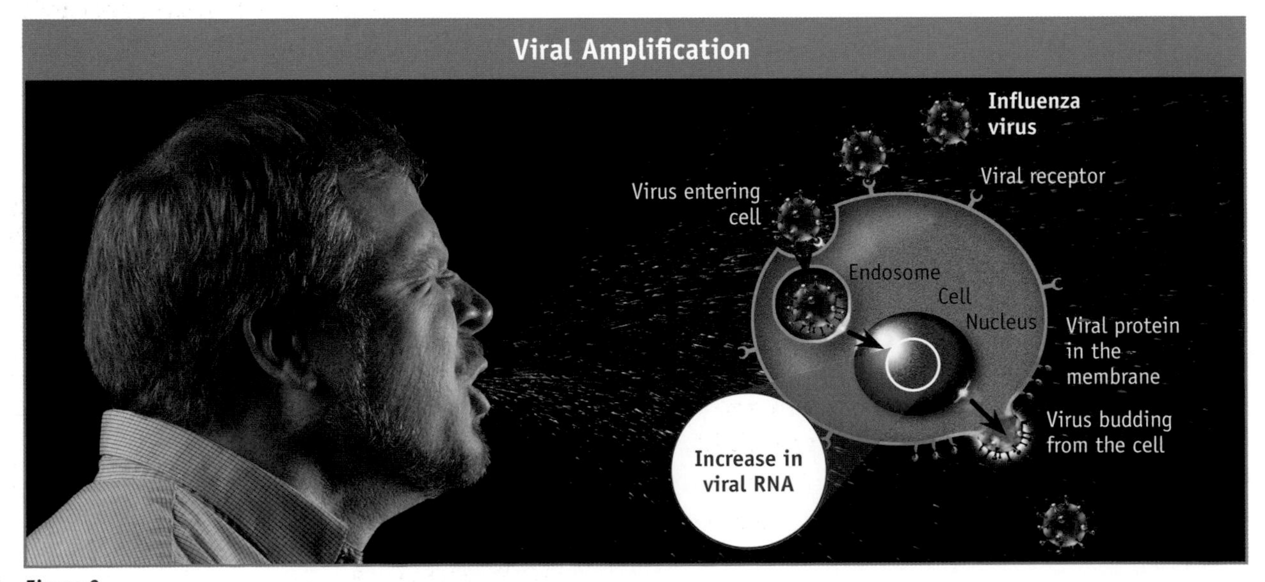

Figure 2

influenza to human beings began approximately 10,000 years ago, when humans began to cultivate land and domesticate animals vulnerable to infection.

Like all viruses, the influenza virus cannot be considered a living organism, since it is unable to reproduce by itself and absolutely has to use a cell to do so. This requirement stems from the fact that viruses live a life of "voluntary simplicity"; they possess only the bare minimum needed to propagate their species. In the case of influenza, this essential material consists of just 11 genes (in comparison, a human being has roughly 25,000) that, together, enable the virus to enter the cells lining the respiratory system and amplify the virus's genes by hijacking the respiratory cells' machinery for its own benefit, so that new viruses can therefore emerge.

Hemagglutinin (H) and neuraminidase (N) are two viral proteins that play an essential role in influenza replication. The wide diversity of type A influenza viruses are due in large part to the variations in one or the other of these two proteins, variations that are now used to describe a strain of virus affecting a given population. For example, when we speak of an H1N1 strain, this means that the virus is made up of a combination of Type 1 hemagglutinin and neuraminidase, whereas an H5N1 strain contains Type 5 hemagglutinin. To date, combinations of 15 types of hemagglutinin and nine types of neuraminidase have been described, chiefly in birds.

These two proteins play a key role in the pathogenicity of the influenza virus. For the virus to manage to infect a cell, hemagglutinin must interact with a receptor on the surface of the cell so that the virus can enter and deliver its genetic material to the nucleus in each cell (see Figure 2). The nature of the interaction between hemagglutinin and its receptor determines the species of animal that can be infected by the virus as well as the degree of contagion or the seriousness of the resulting infection. For example, strains like H1N1, where the hemagglutinin interacts with receptors in the cells of the upper respiratory pathways (nose, mouth, throat), are highly contagious, as the newly formed viruses are easily expelled when a person coughs or sneezes and can therefore infect a new host located nearby.

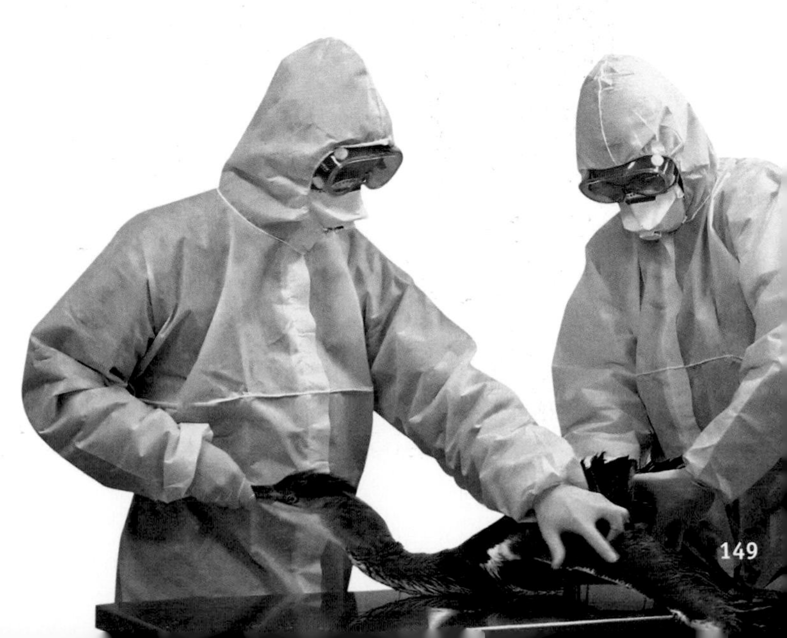

Conversely, some influenza strains originating in birds and containing Type 5 hemagglutinin (H5N1) are not contagious in humans, as they attach themselves to receptors located deeper in the lungs and the new viruses cannot easily escape from the organism (see Figure 3). Despite its low level of contagion, an H5N1 strain currently latent in certain parts of the world can nonetheless be transmitted to humans who have been in direct contact with infected birds. This avian flu then causes sudden viral pneumonia that offers little chance of survival. According to the World Health Organization (WHO), 447 cases of flu caused by the animal-human transmission of the H5N1 virus have been seen in Asia in recent years; 263 of these people died shortly after becoming infected. With a mortality rate of 60%, it is clear that if this virus acquired new characteristics to allow it to spread efficiently between human beings, the consequences could be terrifying.

Since the objective of the virus is to reproduce itself many times over so as to infect the largest possible number of hosts, getting out of the cell is just as important as getting in. This process is more complex than it appears, as the presence of hemagglutinin on the surface of the newly produced viruses means they can be recog-

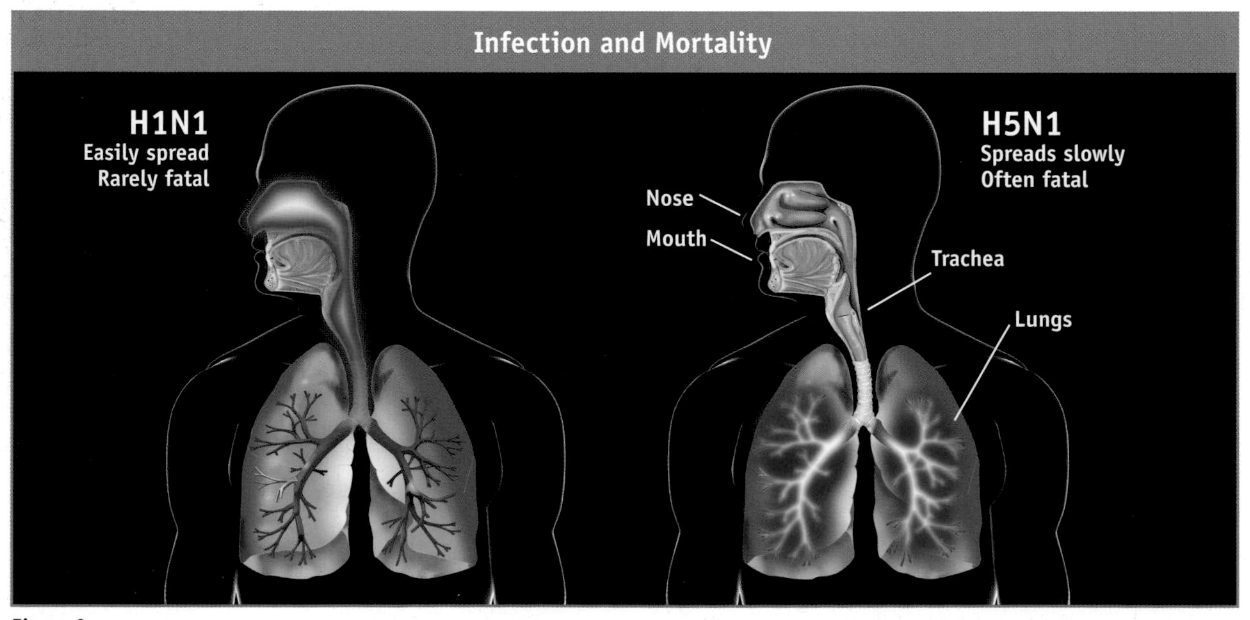

Figure 3

Worldwide flus

Every influenza epidemic in the last century was caused by the sudden appearance of a new strain of Type A virus in animals (birds or pigs) that was subsequently transmitted to humans (see Figure 4).

H1N1 (1918–1920)

The Spanish flu (so-called because the Spanish, who were not involved in World War II, did not try to hide the existence of the disease that was affecting their population and were the first to officially report it) struck a third of the world's population, causing between 20 and 100 million deaths. This epidemic likely stemmed from a mutation in a strain of avian flu (H1N1) that enabled the virus to infect humans, as well as from mutations in certain genes that considerably increased its virulence. The role of this strain was not, however, limited to the 1918 flu, for all subsequent pandemics have been caused by viral strains derived in part from this H1N1.

H2N2 (1957–1958)

The Asian flu epidemic originated in a combination of an avian H2N2 strain (specifically from ducks) with a mutated form of the 1918 H1N1 virus. While less virulent, this strain nonetheless killed more than 2 million people in just two years, particularly in China. The epidemic, which occurred at the same time as the famine caused by Mao Zedong's Great Leap Forward policy, marked an especially dark period in the country's history. However, the H2N2 virus had a short lifespan; the strain had already disappeared and been replaced by its successor (H3N2) just 11 years after it first appeared.

H3N2 (1968–1969)

This highly contagious, but relatively non-virulent, strain was derived from a combination of human and avian viruses. This virus caused the Hong Kong flu in 1968; variants still exist today and are in large part responsible for seasonal flu epidemics.

H1N1 (2009)

This virus is a very complex combination of four distinct strains of influenza originating in pigs, birds and humans. It has received the most media coverage in recent years, but the pandemic caused by the H1N1 strain in 2009 has paradoxically proved to be one of the least dangerous, with a mortality rate roughly three times lower than that of the seasonal flu.

nized by the same receptors that let them in, and they can therefore remain stuck on the cell's surface, unable to make their way to a new host cell. The neuraminidase on the surface of the virus provides a way around this problem, however, by eliminating certain sugars involved in the interaction of hemagglutinin with the receptor. The virions are then able to break the bond with the surface of the infected cell and escape to infect others. The importance of this stage is clearly shown in the antiviral activity of Tamiflu (oseltamivir) and Relenza (zanamivir), two drugs that can end the viral replication cycle by specifically blocking neuraminidase.

An epidemic virus

One of the most remarkable characteristics of the influenza virus is its unmatched ability to change its structure and create new viral types able to circumvent the immune system of the host it is using to replicate itself. Whereas a single vaccine provides life-long protection against poliomyelitis or measles, influenza is caused by a different virus every year, and its prevention requires a new immunization every flu season. The virus's highly changeable nature results from its ability to accumulate random mutations in its protein structure (antigenic drift) that can even, in some cases, give it the ability to infect a new species. Thus, the dreadful Spanish flu epidemic of 1918 was apparently caused by a mutation in an avian virus that gave it the ability to infect human cells. In most cases, however, a new kind of virus is created when distinct influenza viruses simultaneously infect an animal and give rise to hybrid viruses that contain elements from each (antigenic break). For example, the H3N2 virus, currently responsible for the majority of annual flu epidemics, first appeared when pigs were infected with human H2N2 and certain avian viruses, giving rise to a new virus able to infect human beings efficiently.

20th-Century Flu Pandemics

Pandemic	Year	Type of flu	Deaths worldwide	Mortality rate
Spanish flu	1918–1920	A/H1N1	20–100 million	2 %
Asian flu	1957–1958	A/H2N2	1–1.5 million	0.13 %
Hong Kong flu	1968–1969	A/H3N2	0.75–1 million	< 0.1 %
H1N1 flu	2009	A/H1N1	10,000	0.01–0.03 %

Figure 4

This type of recombination is very advantageous for the virus, since it allows it to radically change its infectious potential, or virulence, and thus, with this new profile, bypass immune defenses established during previous infections. When the dice roll in the virus's favor, a particularly virulent and infectious strain is produced, which can trigger an epidemic with fearful consequences.

Every year, it is estimated that influenza strikes between 5% and 15% of the world's population and may cause up to 500,000 deaths, especially among young children, the elderly and those with chronic diseases. It is often forgotten that flu is not a harmless disease!

However, epidemics resulting from the sudden appearance of particularly virulent new strains of viruses are the main threat posed by influenza. In the past 100 years, four main pandemics have infected the world's population, the most disastrous being that of 1918, probably the only epidemic whose devastating consequences can be compared with those of the Black Death.

Dying from the flu

The human body has good resistance to the influenza virus, and in the vast majority of cases healthy people manage to overcome the infection in a few days. However, even though the mortality rate is just 1%, the damage caused by this virus is still significant, given its ability to infect a large portion of the population, as shown by the tens of millions of deaths that occurred during the 1918 epidemic. This is why the emergence of an exceptionally virulent strain such as avian H5N1, potentially able to be transmitted from one person to another, remains a major preoccupation that keeps public health organizations always on alert.

A single virus entering a cell in the respiratory tract triggers a very rapid process: just a few hours after the infection starts, thousands of new viruses have already been produced and are at work infecting adjacent cells. Influenza is a cytolytic virus, that is, when it reproduces, it causes infected cells to die, which triggers a fast response from the organism's immune system and summons inflammatory cells to the site of the cell damage. This inflammation is what actually causes the cough associated with the flu, a reflex whose role is to expel debris and foreign bodies from the respiratory tract. In a healthy person, this inflammation is a call to action that engages the entire immune system and usually results in the complete neutralization of the virus within a few days. On the other hand, when immunity is not at its peak, as in young children, the elderly or ill people, the destruction of respiratory cells by the virus creates an ideal opportunity for other pathogens, especially bacteria found in the respiratory tract. These bacteria can then infect the weakened tissues and cause pneumonia. When the cells that absorb oxygen from the air are affected, pulmonary function may deteriorate and cause death. The vast majority of deaths from the Spanish flu were due to this kind of bacterial pneumonia that was caused by the the influenza virus destroying respiratory cells.

AIDS: The mark of a devastating virus

Acquired immunodeficiency syndrome (AIDS) is the name given to the group of symptoms caused by the destruction of certain immune system cells by the human immunodeficiency virus (HIV). The virus appeared in humans toward the end of the 1970s and is particularly Machiavellian, as it specifically attacks the CD4 lymphocytes, a class of white blood cells that act as conductors for the immune response, coordinating the production of antibodies. By thus disabling the immunity squadron's high command, HIV eventually causes confusion among the soldier cells, which need specific orders to eliminate foreign bodies, and then creates total disorganization once there are no longer enough CD4 lymphocytes. AIDS is the end result of a lengthy process (10 years on average) during which the multiplication of the virus in immune cells slowly but inexorably undermines their function and makes the infected person unable to fight off bacteria, viruses, fungi and other parasites normally handled by our defenses. When this drop in resistance to infections occurs, the disease develops very quickly and usually causes death in less than a year. Most deaths are caused by opportunistic

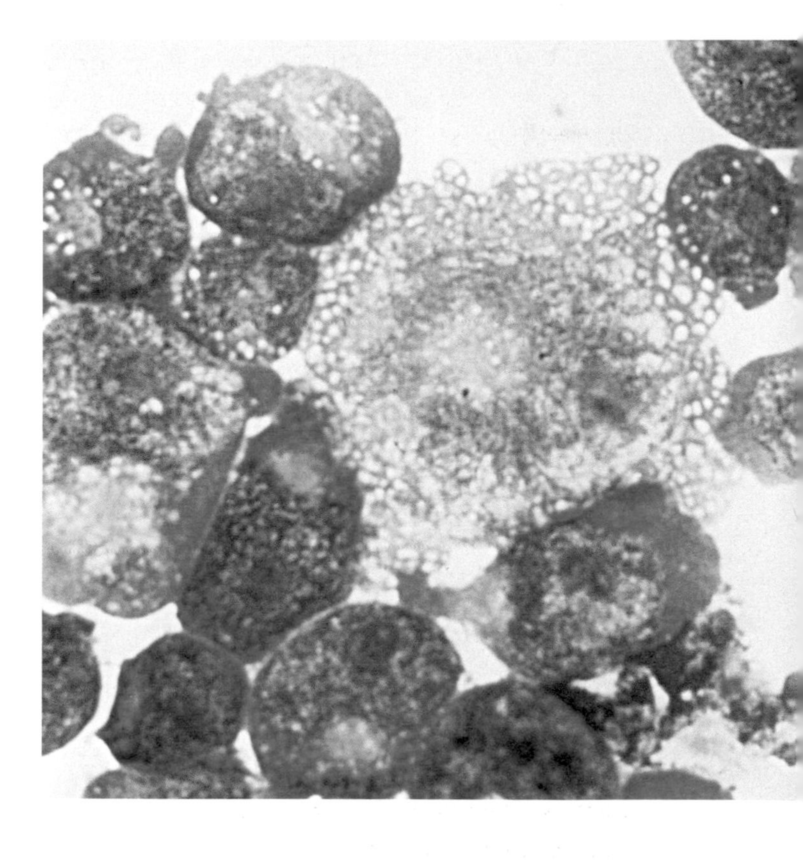

^ Microscopic image of T cells infected by HIV

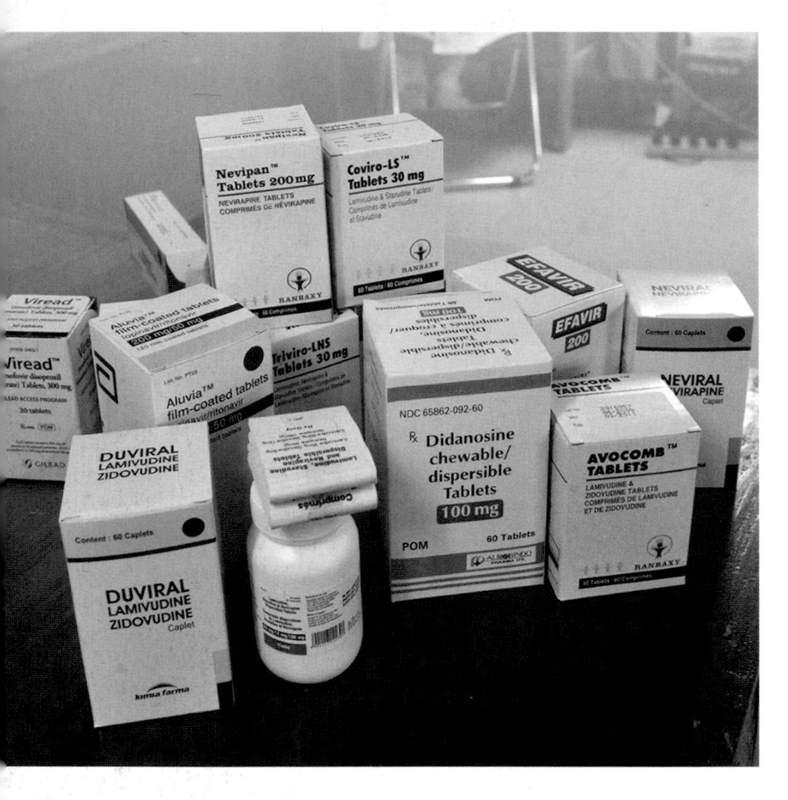

infections, owing to the extreme vulnerability of the immune system: pneumonias, tuberculosis, toxoplasmoses and other infections. Several types of cancer are also associated with the deaths of people suffering from AIDS: lymphomas and especially Kaposi's sarcoma, a rare cancer whose unusually frequent occurrence led to the epidemic's discovery at the beginning of the 1980s. Even though in industrialized countries the death rate associated with AIDS is not as nearly as high as for chronic diseases (for example, in the United States 20,000 deaths were attributed to AIDS in 2007 compared with 550,000 deaths from cancer), the disease has struck certain African countries particularly hard, where it is now the main cause of death.

The extraordinary contribution of scientific geniuses like Louis Pasteur (asepsis, vaccines) and Alexander Fleming (penicillin) have played a leading role in the dramatic drop in mortality associated with infectious diseases. In spite of this, microbes continue to be a constant threat to human life. Whether it be the recent appearance of the AIDS virus, the strong resurgence of bacteria resistant to antibiotics, the emergence of absolutely terrifying viruses like Ebola and Marburg or even the strong likelihood that a new strain of virulent flu will appear in the coming years, infectious agents remind us that the fight against these invisible enemies has not been won and likely never will be. Our inborn fear of microbes is therefore justified: no disease, no poison and no weapon, no matter their strength or range, has a destructive potential that matches that of microbes, able to kill in record time hundreds of millions of people. Will microbes one day indeed have the last word?

^ Some of the antiretroviral drugs used to treat HIV.

> Bayo Iribhogbe, *Untitled II – African Aids Epidemic*

Chapter 7

Poisons: Fascination and Danger

– Winston, if you were my husband, I'd put poison in your tea.
– Madame, if you were my wife, I would drink it.
Winston Churchill (1874–1965) in response to Nancy Astor (1879–1964)

Nearly 3,000 years ago, Phoenician merchants living in Sardinia devised an extremely macabre ritual for putting to death elderly people unable to take care of themselves. First, they were poisoned with a paralyzing potion, and then they were either thrown off the top of a cliff or stoned or beaten to death. Strangely, the contraction of the facial muscles caused by the poison gave the dead person a grimace that looked like a smile, almost as if the person were expressing gratitude for being, at last, relieved of the burden of existence. The reasons behind this, to say the least, radical euthanasia remain unclear, but these "smiling dead" have nevertheless left their mark. When Homer wrote "Ulysses turned his head a little aside, and avoided it, smiling grimly Sardinian fashion as he did so." (*The Odyssey, Book XX*), he was making a reference to a slightly mocking or sardonic smile, like that of the aged of Sardinia "who had lived long enough."

We now know that the famous potion was made from hemlock water dropwort (*Oenanthe crocata*), a plant of the Apiaceae family whose roots have a mild and pleasant taste that is a little like turnip. The tubers, however, are extremely poisonous, as they contain a powerful neurotoxin called oenanthotoxin. This neurotoxin causes spasms in the facial muscles and the appearance of what we call a *risus sardonicus*, or grimace, owing to the contraction of these muscles. (We might wonder whether the effects of this plant didn't inspire the creators of the Joker, Batman's sworn

< Giovanni Pedrini Giampietrino, *The Death of Cleopatra* (detail)

enemy, whose preferred deadly weapon is a poisonous substance that causes a smile to appear on his victims' faces.)

Sardonic toxin is very nearly identical to cicutoxin, one of the toxic molecules found in poison hemlock (*Conium maculatum*). This well-known plant was one of the principal ingredients in the poison used in ancient Greece to execute those condemned to death. Notably, it was drunk by Socrates, accused of having "corrupted" Athenian youth with his ideas. These two plants, while completely innocent looking, are thus a good illustration of the immense dangers hidden in the natural world around us and, above all, of the ability of humans to recognize these poisons and use them for murderous ends.

Dangerous plants

Although we often think of chemical warfare as a recent threat resulting from the undesirable effects of an ever more destructive technology, this strategy is in fact the carbon copy of a tactic perfected by plants several million years ago. Facing a threatening danger, animals (including humans) spontaneously adopt one of two strategies: either direct combat to neutralize the threat or flight, in the hope of avoiding the unfortunate consequences of a confrontation with a more

^ Hemlock plant

Toxic Plants

Castor oil plant

An ornamental plant greatly admired for the majestic appearance of its palmate lobed leaves, the castor oil plant (*Ricinus communis*) is mainly cultivated for the many industrial applications of the oil in its seeds. Ricin oil (formerly known as "castor oil" because it replaced castoreum, which is secreted by the sex glands of beavers) is harmless and was long used in folk medicine as a laxative as well as to trigger uterine contractions. However, the seeds of the castor oil plant also contain a large quantity of ricin, a protein that is extremely toxic (roughly 6,000 times more toxic than cyanide) and can completely block the synthesis of proteins in cells. The enormous toxicity of ricin was particularly well demonstrated in the dramatic assassination of dissident Georgi Markov by the Bulgarian secret service, on September 17, 1978. A fierce critique of the Bulgarian communist regime since leaving the country in 1969, Markov was crossing the Thames on London's Waterloo Bridge when he was jabbed in the leg by a man carrying an umbrella. He developed a high fever that same night and died 3 days later. The autopsy revealed the presence of a tiny platinum pellet the size of a pinhead in his calf along with traces of ricin on the surface of the object. According to the data available, the murder is thought to have been committed by Francesco Giullino, an agent of the Durzhavna Sigurnost (the Bulgarian equivalent of the KGB), using an umbrella fitted with a pneumatic pellet-shooting mechanism.

Oleander

In spite of its beauty and the elegance of its flowers, oleander (*Nerium oleander*) is one of the most toxic plants on earth. All of its parts contain very powerful poisons, notably oleandrin, neriantin, nerianthoside and rosaginoside. These alkaloids interfere with the activity of ATPase Na^+-K^+ in the heart muscle cells, preventing muscular contraction and hence causing the heart to stop. The toxicity of these alkaloids is such that a single leaf of oleander can kill a child.

Yew

A shrub frequently grown for ornamental purposes, the common yew (*Taxus baccata*) is nonetheless

one of the most toxic trees on earth. All parts of the plant (except the flesh of the arils in female plants) contain taxin, a complex mixture of alkaloids that are rapidly absorbed in the intestines and cause the heart to stop beating. The yew's high toxicity has been known since time immemorial, with concoctions of the plant even being used by desperate women to induce abortion; unfortunately, the death of the mother often preceded that of the fetus. However, the yew has more than just negative qualities: the bark of a North American species (*Taxus brevifolia*) contains a molecule called Taxol that is used in chemotherapy for uterine and breast cancers (paclitaxel). Since the treatment of a single patient calls on average for the bark of five or six 100-year-old trees, other ways have been found to synthesize the drug from the common yew's needles. This has led to the discovery of Taxotere, a molecule twice as effective as Taxol in fighting certain tumors.

powerful adversary. Less evolved organisms such as plants, however, are not able to use this fight-or-flight strategy. They cannot pull up their roots and run for their lives to escape from a threat, and their lack of nervous and muscular systems obviously rules out any possibility of a physical fight with their aggressors. This is a major problem for organisms at the bottom of the food chain! To get around these limitations and avoid being completely destroyed by herbivore animals or plant-eating microorganisms (viruses, bacteria, fungi), plants have, in the course of their evolution, developed a staggering array of highly toxic molecules — devastating poisons that can rapidly cause the death of the unwary would-be threat.

There is therefore nothing out of the ordinary about the powerful toxicity of dropwort and hemlock mentioned earlier. The presence of deadly poisons is, on the contrary, an intrinsic characteristic of the great majority of plants, even several magnificent ornamental plants commonly grown in our gardens (see box on pp. 161–162).

The toxicity of these molecules is even used in modern pharmacology, and about half of the drugs used in chemotherapy for cancer have plant origins. The beauty of plants is often only equaled by their devastating toxicity!

This extremely efficient strategy has had extraordinary impacts, not only on the biodiversity of the plant world, by giving plants a broad ecological niche, but also on all life on earth, which has had to adapt to the fact that poisons are everywhere. The strategy used by the monarch butterfly (*Danaus plexippus*) is an excellent illustration of how ingeniously some plants and animals have evolved: in the larval phase, the monarch caterpillar feeds on common milkweed, a plant full of cardiac poisons (cardenolides) and stores these poisons in a specialized compartment until it turns into an adult butterfly. The concentrations of poisons inside the butterfly are such that it causes vomiting in the birds that try to put them on their menu, thus effectively protecting the insects. This is a very complex evolutionary process, and some species of birds have in turn acquired resistance to the monarchs' poisons! Thus, as paradoxical as it may seem, the mind-boggling diversity and beauty of nature are very often the direct consequences of the "cold war" fought between many predators and their prey, each held in check by the toxic power of its counterpart. This ongoing chemical warfare is clearly shown by animals having evolved to produce a large number of enzymes, particularly the cytochromes P450, which can transform these toxic molecules into less dangerous substances that the organism can eliminate before they cause too much damage. In human beings alone, no fewer than 57 genes are dedicated to producing various cytochromes P450!

Deadly bites

Although the vast majority of poisons are found in the plant world, several animals have also developed toxic arsenals that are just as deadly as those of plants. A considerable number of species began to produce poisons in the form of venom very early in evolution (see box on pp. 164–165).

Venoms do not owe their toxicity to molecules with relatively simple structures, as is the case with plant poisons, but to complex mixtures of proteins that target several essential life processes. The venom of certain snakes, for example, is an extremely complex mixture that may contain many different enzymes meant to facilitate the digestion of prey, along with hundreds of toxins that paralyze the respiratory system and stop the heart from beating or attack the blood vessels and muscles causing hemorrhaging and necroses (see Figure 1).

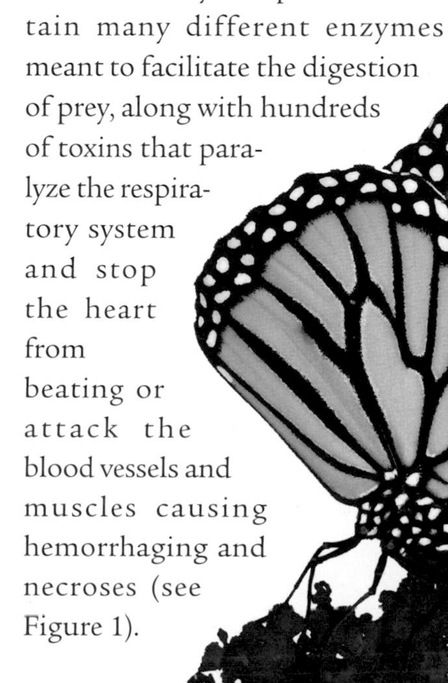

^ Monarch caterpillar

> Monarch butterfly

Toxic animals

Geographic cone snail

While most of the roughly 500 species of sea snails (also called cone snails) are harmless, some members of this family are formidable carnivorous predators. The most dangerous species, the geographic cone snail (*Conus geographicus*), feeds on fish using a sharp proboscis armed with a harpoon-like tooth coated with venom containing a large range of tiny proteins, called conotoxins, that block the nerve impulse, hence paralyzing the victim. Its venom is one of the most poisonous in the world, capable of killing a person in 2 hours.

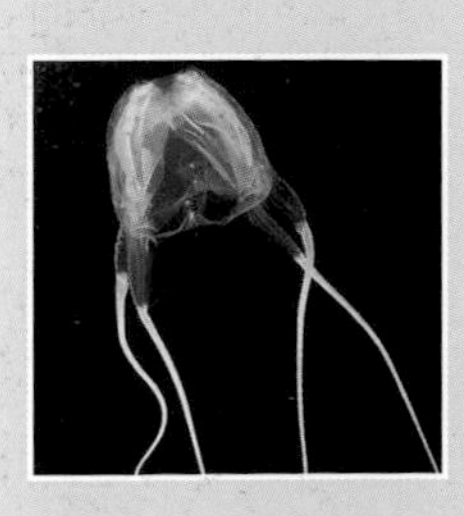

Box jellyfish

The Australian box jellyfish, (*Chironex fleckeri*), also called the sea wasp, is one of the world's most poisonous marine animals. Its numerous tentacles, that can reach lengths as much as 13 feet (4 m), each contain 500,000 cnidocytes, poisonous cells equipped with a harpoon that can become lodged in the victim's skin after contact with the jellyfish. The venom contains many proteins that cause dreadful pain for several weeks and, at high doses, paralyze the heart and lungs causing death in under 5 minutes. Since each tentacle contains enough toxins to kill 50 people, it is not surprising that this jellyfish is the principal cause of death from a marine organism, responsible for more than 5,000 deaths since 1954.

Golden poison frog

Despite its small size (just over 1 inch/35 mm), the golden poison frog (*Phyllobates terribilis*) is currently considered to be the most poisonous vertebrate in the world. Native to Colombia, one frog alone exudes enough batrachotoxins through the pores of its skin to kill 20,000 mice or 10 humans!

Desert taipan

The taipan (*Oxyuranus microlepidotus*) has the most toxic venom in the snake family. A single dose can kill 100 people! Fortunately, the taipan lives in the arid regions of central Australia and is, by nature, rather unsociable.

Deathstalker scorpion

Found mainly in the deserts of North Africa and the Middle East, the deathstalker (*Leiurus quinquestriatus*) is by far the world's most dangerous scorpion. Straw yellow in color and between 3½ and 4½ inches (9–11.5 cm) in length as an adult, this arthropod produces a toxin that interfers with the entry of chlorine into the neurons, blocking the nerve impulse and causing death by paralysis.

Snakebites Are Destructive

Class	Examples	Mechanisms of action
α-neurotoxins	α-bungarotoxin, α-toxin, erabutoxin, cobratoxin	Neurotoxins block neuromuscular transmission by bonding, like curare, with the cholinergic receptor in skeletal muscle fibers.
κ-toxins	κ-toxin	κ-toxins block certain cholinergic receptors in the central nervous system.
β-neurotoxins	Notexin, ammodytoxin, β-bungarotoxin, crotoxin, taipoxin	ß-neurotoxins block neuromuscular transmission by preventing nerve endings from releasing acetylcholine. May interact with a voltage-sensitive potassium channel.
Dendrotoxins	Dendrotoxin, toxins l and k	Dendrotoxins increase the amount of acetylcholine released by nerve endings. May interact with a voltage-sensitive potassium channel.
Cardiotoxins	γ-toxin, cardiotoxin, cytotoxin	Cardiotoxins disturb the plasma membranes in certain cells (e.g., heart fibers, excitable cells) and cause them to rupture (lysis). They cause cardiac arrest.
Sarafotoxins	Sarafotoxins a, b and c	Sarafatoxins a, b and c are powerful vasoconstrictors that affect the entire cardiovascular system. They cause cardiac arrest.
Myotoxins	Myotoxin-a, crotamin	Myotoxins cause the degeneration of muscle fibers by interacting with a voltage-dependent sodium channel.
	Phospholipase A2	Phospholipase A2 causes the degeneration of muscle fibers.
Hemorrhagins	Mucrotoxin A, hemorrhagic toxin a, b, c, HT1, HT2	Hemorrhagins cause very significant hemorrhaging by changing the blood vessel walls.

Figure 1

Source: R. Bauchot. *Serpents* (2005).

The deadly power of snake venom is especially strong in the members of the Elapides (taipans, cobras, mambas, najas, sea snakes and coral snakes) and the Viperides families (in particular Russell's viper, which is highly irritable by nature). Among the approximately 5 million people bitten each year by snakes, those who have the bad luck to run into one belonging to these species risk joining the ranks of the 120,000 deaths annually caused by snakebites.

Tamed poisons

Detecting the presence of poisons to avoid death is obviously a basic skill for the survival of a species, a fact which is clearly illustrated by the large quantity of genes associated with the senses of taste and smell in most living species. In human beings, detection also relies on cultural input, since knowledge acquired about the toxic properties of a plant or animal is transmitted from generation to generation. While this knowledge is of primordial importance to identify foods that can safely be consumed, it is also true that the knowledge acquired about the toxic properties of several elements in nature has opened up new possibilities with regard to the use of these poisons. No doubt one of their early uses was to improve hunting methods. For example, more than 18,000 years ago the Masai hunters of Kenya were using plant extracts containing powerful cardiotoxins to increase the effectiveness of their arrows, a strategy similar to that of the peoples of South America who used curare. Unfortunately, it was but a short step for humans to begin using poisons' deadly effects on other humans. The new weapon was soon used for the purposes of war, as shown in the origin of the word toxic, which is derived from the Greek *toxicon*, meaning "poison for an arrow."

Far from being restricted to arrows, the use of poisons for murderous ends seems to be as old as human civilizations themselves. The

< Cobra in attack position.

^ A Yanomami man drying poison on arrows by the fire.

Codex Ebers, a 3,500-year-old Egyptian medical text, describes the toxic properties of a whole range of substances, in particular arsenic, mandrake, hemlock and aconite. The ancient Egyptians had even discovered a way of extracting a powerful poison from peach pits, which they used to subject people suspected of crime to "trial by poison." According to the beliefs of the time, these substances were lethal to the guilty but harmless to the innocent. As a result of this "punishment by peach," many innocent people must have been unjustly condemned, for we now know that this extract contains amygdalin, a molecule that releases cyanide on its way through the intestine! The use of a poison as dreadful as cyanide so long ago did not bode at all well for the inventiveness of human beings in this field... And indeed, whether used as instruments to gain power or to extract vengeance, poisons quickly became substances that were practically inseparable from humanity's criminal impulses.

Attacking our weaknesses

Since oxygen is essential to life, the most efficient means of causing death is obviously to cut off a living being's oxygen supply as quickly as possible. Poisons are particularly efficient for this, given the incredible speed with which they can interfere with the transformation of oxygen into energy directly at the cellular level or indirectly, by blocking the organs that play an important role in blood circulation and hence that of oxygen (e.g., the lungs and heart).

Cyanide and arsenic Cyanide and arsenic are undoubtedly the best-known poisons, both because of their classic role in crime literature and their repeated use for lethal purposes throughout history. During the Roman Empire, a period when poisonings were commonplace in the circles of power, cyanide was very useful to Nero in eliminating his rival Britannicus (aided by Locusta, a famous Roman female poisoner). In the Middle Ages, a time when poisons were one of the surest and fastest ways to climb the social ladder, the trio made up of Pope Alexander VI (Rodrigo Borgia) and his children Cesare and Lucrezia turned the systematic poisoning of rivals into an almost recreational activity. They used a mysterious mixture called cantarella (trioxide of arsenic and phosphorus), capable of either killing

< Salvator Rosa, *Death of Socrates* (detail)

instantly or over several days, depending on the dosage. It is no surprise that, even today, the Borgia name remains synonymous with intrigue and murder!

Arsenic's effectiveness has been confirmed many times over. Some preparations, like the aqua toffana of Madame Giulia Toffana and the "succession powder" of Madame Deshayes (known as "The Neighbour"), were used on large scale to eliminate many undesirables, including a number of troublesome husbands.

This "popularity" of cyanide and arsenic as instruments of murder is largely due to the fact that these substances act directly on the mechanism responsible for transforming the oxygen we breathe into ATP, the biochemical energy cells can use (see Chapter 2). Cyanide, for example, bonds very quickly to cytochrome c oxidase, a very important protein that is part of the respiratory chain involved in ATP synthesis. By binding with the iron in this protein, cyanide blocks the chain at this level and stops ATP from being produced. The cell then quickly becomes unable to "breathe" properly, which means that the organs that depend on a constant supply of oxygen, such as the brain and heart, stop functioning. Cyanide is particularly dangerous in its gaseous state, hydrogen cyanide, which is a very toxic compound that directly attacks the lungs. It was the active ingredient in the notorious Zyklon B, the product used by the Nazis in the death camp gas chambers during the Holocaust.

Aside from the many murders committed using cyanide, the deadly power of this molecule is also clearly seen in a number of famous suicides, notably that of Hermann Göring, the Nazi Luftwaffe commander who swallowed a potassium cyanide capsule the night before his execu-

^ Portrait of Rodrigo Borgia, who was Pope Alexander VI, by an unknown artist.

^ One of the gas chambers in the Nazi concentration camp at Dachau, Germany.

tion for war crimes, as well as in the mass suicide of 909 followers (276 of them children) of Jim Jones's People's Temple on November 18, 1978, in Guyana.

However, poisoning is not always deliberate. More than 70 million people in Bangladesh have been exposed to toxic concentrations of arsenic in drinking water, making it the largest poisoning in history according to the World Health Organization (WHO). By a cruel stroke of bad luck, this catastrophe is the unfortunate consequence of measures recommended by development organizations in the 1970s and 1980s to encourage people to stop drinking surface waters, which are known to harbor diseases such as cholera. More than 10 million wells were dug, but only recently did it come to light that a high percentage of them (40%) have arsenic concentrations that may be more than 10 times the maximum allowable quantity recommended by the WHO. According to experts, prolonged exposure to these concentrations of arsenic may be responsible for 125,000 cases of skin cancer and 3,000 deaths caused by cancers affecting internal organs.

Poison gases: carbon monoxide Carbon monoxide (CO) is another form of metabolic poison that has an effect on oxygen supply to the cells. This gas is particularly dangerous, as it is colorless, odorless, and tasteless, and it takes only very weak concentrations of it to cause death. Carbon monoxide poisoning is in fact the most common form of poisoning in most industrialized countries.

CO is produced by the incomplete combustion of a carbon source, whether in the form of a hydrocarbon (gas and petroleum derivatives) or organic matter (wood

^ A container of Zyklon B gas used as a deadly poison in the gas chambers in a number of Nazi death camps.

or charcoal). Although long known to be dangerous (the Greeks and Romans used charcoal's toxic smoke to execute criminals and commit suicide), it was only thanks to the work carried out by Claude Bernard, the great French physiologist, that the way it works was explained.

We now know that the toxicity of CO stems mainly from its ability to interact with hemoglobin, which transports oxygen to the cells. Each molecule of hemoglobin has four binding sites for oxygen, which work together to maximize the capture of oxygen in arterial blood flowing from the lungs and release it into the tissues and organs. In the presence of carbon monoxide, however, this cycle is completely disrupted, as this toxic gas has an affinity for hemoglobin that is 200 times greater than oxygen, and hence stops the carrier from efficiently capturing oxygen. This binding of carbon monoxide is all the more dangerous because it inhibits the release of oxygen molecules found on hemoglobin's other binding sites. Consequently, even if the concentration of oxygen in the blood increases, this oxygen remains bound to the hemoglobin, unable to be delivered to the cells. This situation throws the organism into real panic, as the heart tries to compensate for oxygen deprivation by beating much faster (tachycardia), increasing the risk of angina, arrhythmia and pulmonary edema. The brain, an organ extremely dependent on oxygen, is also an early target for the toxic gas, and headaches, nausea and convulsions are classic symptoms of this type of poisoning. If this situation is not quickly remedied by providing a massive supply of oxygen to displace the carbon monoxide, the cutting off of the oxygen supply results in death. All activities involving a process of combustion (automobiles, oil or gas stoves, etc.) can produce CO, and it is vital to make sure the fumes thus released are not confined to an enclosed space.

Nervous movements

Acetylcholine, a neurotransmitter synthesized inside neurons and stored near the neuromuscular junction, plays an absolutely essential role in the response of muscles to a signal from the nervous system. When a nerve receives a stimulus, acetylcholine is released and spreads very quickly to interact with receptors on the surface of the muscle cells. This signal triggers a complex chain reaction that culminates in the contraction of the muscle fibers. However, the neurotransmitter must be eliminated from the synaptic space so the next wave of nerve impulses can, in their turn, stimulate the muscle. This elimination is made possible by acetylcholinesterase, an enzyme occurring in the synaptic space that destroys acetylcholine. The importance of this process is well illustrated by the evolutionary perfection of acetylcholinesterase, which can destroy close to 4,000 molecules of acetylcholine per second. Consequently, the lifespan of acetylcholine at the neuromuscular junction is only about 1 or 2 milliseconds.

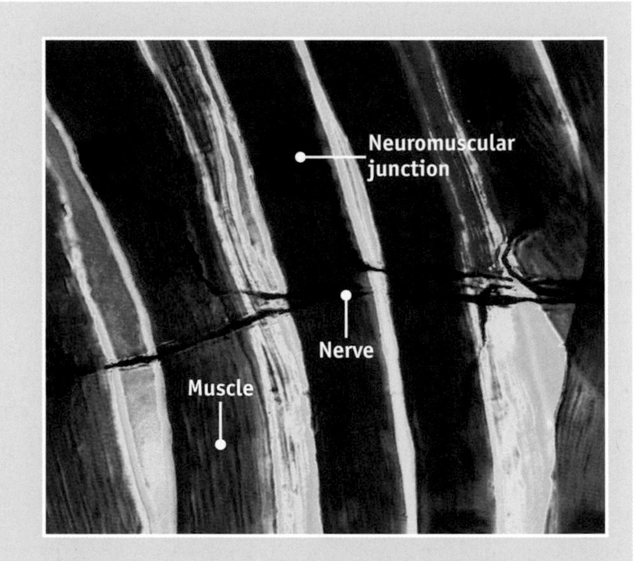

^ Neuronal axon endings at a neuromuscular junction.

Neurotoxic poisons: Chemical warfare, strychnine, hemlock, curare Whereas poisons targeting mitochondria kill by their direct action on how cells use oxygen, other types of poisons act in an indirect way, by preventing oxygen from reaching the cells. This method is more complex but just as effective!

These poisons have the distinctive ability to interact directly with certain components of nerve cells and to inhibit the normal transmission of nerve signals. In some cases, this blockage occurs at the neuromuscular junction, the place where a nerve impulse is decoded by the muscles to cause a contraction and movement (see box on p. 172). Nerve gases (sarin, tabun, V series gases) and some organophosphorus pesticides (malathion, parathion) are undoubtedly the best examples of poisons that exercise their toxic effect by targeting this process. These neurotoxic agents bind with acetylcholinesterase and prevent it from decomposing acetylcholine, resulting in an accumulation of the neurotransmitter at the neuromuscular junction. The constant stimulation of receptors caused by the surplus acetylcholine triggers violent muscular spasms followed by paralysis of the diaphragm muscle, which quickly leads to respiratory failure and death by asphyxiation. The catastrophic impact of these acetylcholinesterase inhibitors is well demonstrated by the effect of sarin (nerve gas), a molecule 500 times more toxic than cyanide and capable of killing a person in less than a minute.

In total contrast to acetylcholinesterase inhibitors, some poisons, like the curare used by South America's indigenous people or certain toxins derived from snake venom, block acetylcholine from binding with its receptor and thus inhibit the transmission of all nerve impulses. The immediate effect of this action is the paralysis of the diaphragm and respiratory failure.

Finally, a large array of poisons cause death by directly targeting nerve cells to inhibit their transmission of the electric impulses several organs require to function. One of the best examples of this mechanism is the horrible death caused by strychnine poisoning, an alkaloid extracted from the fruits of the nux vomica tree (*Strychnos nux-vomica*) used as rat poison (and sometimes to cut street heroin). Strychnine interferes with the signal produced by glycine, an inhibitory neural

^ Drawing of *Strychnos nux-vomica*, the plant that produces strychnine, from *Phytographie médicale* (Medical phytography) by Joseph Roques.

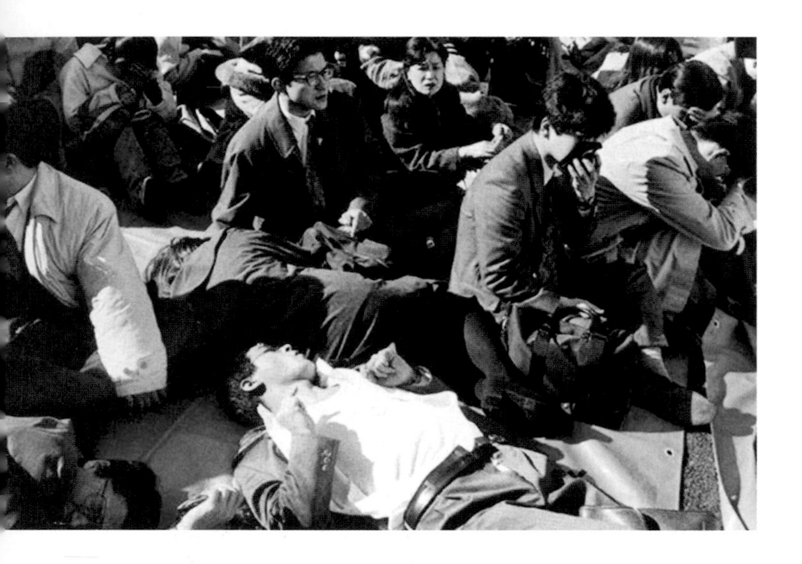

transmitter acting on several regions of the brain to prevent uncontrollable muscle spasms. About 10 minutes after ingestion, the loss of muscle control caused by the poison manifests itself in ever more intense spasms that can increase to the point of causing opisthotonos, a rigid state where the body arches backward, and, in some cases, a *risus sardonicus*. Between 2 and 3 hours after the onset of the effects, the muscles are exhausted and the paralysis of the respiratory pathways causes death.

Food poisoning Even when care is taken to avoid toxic plants and animals, certain kinds of food contamination by microorganisms can cause serious food poisoning. More than 200 diseases are known to be transmitted by food, and in the United States these bouts of contamination cause some 9,000 deaths annually. In spite of their rarity and the relatively low number of deaths attributed to them, they receive significant media coverage and deserve our attention.

Food poisoning is not a recent phenomenon. Historically, one of the most dramatic examples of death from tainted food is ergotism, a disease caused by a fungus (*Claviceps purpurea*) that grows on rye grains. This fungus produces ergotamine, a very toxic alkaloid that causes, in people who have eaten contaminated rye, intense heartburn, hallucinations, convulsions and even the loss of limbs, owing to the significant reduction in blood circulation caused by the poison. (In the Middle Ages, several tales were told of how people with this disease went to see the doctor carrying with them the arm or leg they had just lost.) Some scientists have argued that the symptoms associated with this disease, known as "St. Anthony's fire," strongly resemble those described in the Salem witch trials. They have raised the interesting possibility that the symptoms of bewitchment observed in the young women in Salem were related to rye contamination, since rye was a cereal widely cultivated in the region. Fortunately, ergot poisoning in rye has become so rare as to be almost nonexistent and has now been replaced by other microbial food contamination, due mainly

^ On March 20, 1995, members of the Aum Shinrikyo sect caused chaos when they attacked the Tokyo subway system with sarin gas at rush hour, killing 12 people and injuring more than 5,000.

to faulty food preservation procedures.

Botulism Botulism is a rare but serious food poisoning that is caused by the presence of botulism toxins in canned foods. These toxins, produced by the *Clostridium botulinum* bacterium, are the most powerful poisons in the living world—just one microgram (or a 28 millionth of an ounce) can kill a human being. After contaminated food has been eaten, the toxins penetrate into the neuronal cells and destroy a number of proteins that are essential for the release of acetylcholine at the neuromuscular junctions. The absence of a neurotransmitter inhibits muscle contraction and leads to respiratory paralysis and hence to death. Since *C. botulinum* is a bacterium that cannot survive inside the human body, botulism is always caused by toxins originating in food, generally in home-cured meat or inadequately sterilized canned food. Luckily, the bacterium is very sensitive to heat and can be destroyed simply by boiling.

More recently, botulism toxin has acquired a greatly enhanced reputation as a result of its use as a cosmetic compound (Botox): injected locally, the poison causes paralysis of the muscles underlying wrinkles, smoothing away lines for several months.

Deadly hamburgers *E. coli* O157:H7 is a bacterium found in its natural state in cattle intestines. It makes its way into human food when animal carcasses are contaminated in slaughterhouses or manure containing the bacterium is used as a fertilizer when growing certain vegetables. The food most susceptible to contamination by this bacterium is ground beef. Contamination occurs when the meat comes into contact with the animal's internal organs or excrement when the animal is being slaughtered or the carcass is being butchered. Although all the cuts from a contaminated carcass may contain the bacteria, these remain on the meat's surface and can therefore be destroyed by cooking the meat at high temperatures. With ground beef, on the other hand, the bacteria are found all through the meat and can survive if the meat is undercooked.

While this bacterium is completely harmless for cattle, given its inability to attach itself to and penetrate the cells of these animals, in the human body it can cause serious complications, which first appear as abdominal cramps and bloody dysentery. In some people, young children and the elderly in particular, the bacteria can take

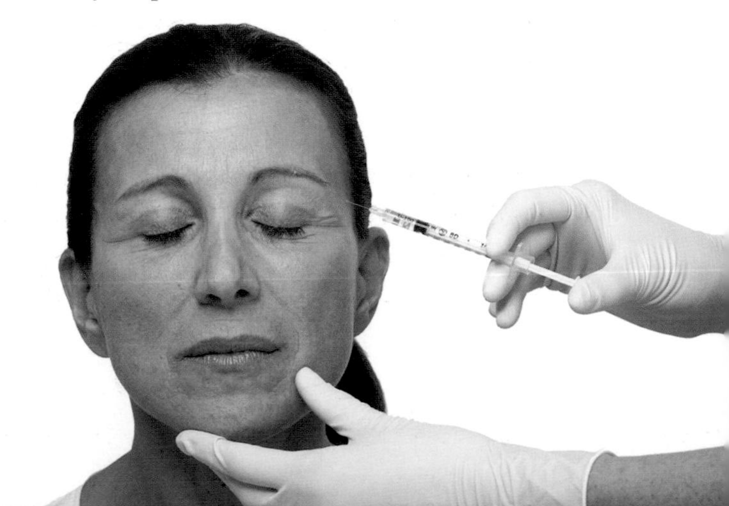

hold and produce a powerful toxin (shiga toxin) that gets into the blood and attacks the blood vessel walls. This causes a complex cascade effect leading to the development of hemolytic uremic syndrome. This disease is characterized by a significant drop in the number of blood platelets (thrombocytopenia), the destruction of red blood cells and kidney failure. At a certain point, organ damage becomes irreversible, causing death.

Deadly poisons can thus be inorganic, plant-based or originate in animals. They have in common the ability to target the two weak points in our biochemical evolution: an absolute dependence on oxygen for metabolic energy and an extremely sophisticated coordination of neurotransmitters enabling our brain to control our whole body. The biological world's evolutionary strategies of attack or defense are thus primarily oriented toward these ideal targets.

Plant poisons are, above all, defensive in nature and serve to protect the plant, which is thus inedible. These toxins are the plant kingdom's most important defense mechanism. The fact that poisons are found throughout the plant kingdom and the importance of plant-based poisons throughout human history is still reflected in our strong sensitivity to bitterness, a chemical characteristic shared by many of these poisons. This taste adaptation has saved humans from poisoning many times in the 200,000 years we have been gathering and tasting the 400,000 species of plants on earth, out of which we have selected the 25,000 species we call fruits, vegetables, herbs, spices, tea or chocolate...

Animal poisons, on the other hand, can be either defensive or offensive. As a defensive tactic, their presence is often associated with vivid colors, which are a sign of danger for predators, which prefer to avoid them. As an offensive tactic, attack molecules serve to paralyze or kill prey, making it easier to capture and digest it. Poisons are a remarkable example of the incredible complexity of adapting to life on earth. They have been, for good reason, the object of veneration or fear in every culture and every era.

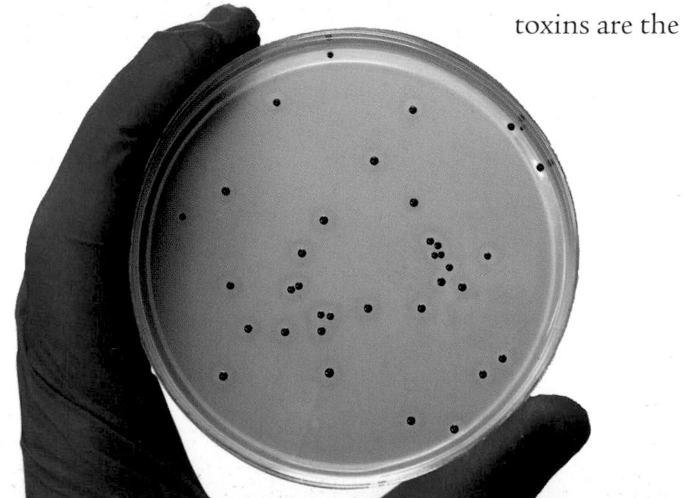

< A golden staphylococcus culture (*Staphylococcus aureus*), the bacterium responsible for certain infections.

< A *Phyllobates terribilis* frog—just touching it can kill a human being.

Chapter 8

Violent Deaths

> There is no beast in the world so much to be feared by man as man.
>
> Michel de Montaigne (1533–1592)

The role of the massive release of adrenalin associated with the fight-or-flight response, described earlier, is to prepare the individual to respond to a threat, either by fighting back or by running away. A human being does not, however, have sharp claws or teeth to attack with, nor are we especially agile, protected by a thick skin or shell, or quick enough to escape from predators. Our ability to survive the pressures exerted on us by other living beings in the process of natural selection and evolution is not due to any outstanding physical characteristic, but rather to the significant development of our cerebral cortex, which gave us survival tools designed to make up for our weak anatomical defenses. In a world where the strongest is generally the one who succeeds in imposing the law, the domination of nature by an animal species as weak as our own is truly an exception, an anomaly that makes us the most unusual animal ever to have lived on the planet.

In a famous scene from the movie *2001: A Space Odyssey*, a tribe of starving prehistoric men, chased away from their watering hole by a rival group, comes up with the idea of using bones as weapons to kill the game they need for survival. Saved by this key discovery, they do not, however, stop there; they use these weapons to attack the enemy group at the watering hole and kill their chief. This scene is interesting because it shows the extent to which violence is inextricably linked with humanity's evolution. In fact, whether for food, to reproduce or to take a rival's possessions, the discovery or invention of new weapons able to generate superhuman strength, subdue adver-

< Bones of the victims of Pol Pot's Khmer Rouge.

saries and impose a particular world view has always been one of the main forces behind innovation in human societies. A number of materials (Teflon, Kevlar, etc.) and most of the technologies that have become essential to modern life (lasers, computers, Internet, etc.) are the fruits of work first carried out for military purposes and that might never have taken place without the energy and enormous financial support dedicated to the development of new weaponry.

The arms race

The common principle underlying all weapons is to neutralize or kill as quickly as possible while minimizing the risk of being hit in return. In this sense, despite their primitive nature, prehistoric clubs—artificial extensions of the arm—marked one of the most important stages in the transition from the big monkeys to the first hominids; for the first time, it was possible to greatly increase strike force. The effectiveness of this close quarter combat was later enhanced by the creation of the first knives in stone and flint, which are the ancestors of the large family of bladed weapons (knives, sabers, swords, etc.). The basic principle of these weapons is to concentrate all the strength onto a very small surface (the tip or sharp edge of the blade) to make it easier to drive the weapon into the adversary's body. Just how dangerous a bladed weapon is must not be underestimated: for example, whereas a bullet shot from a gun has, on contact with the skin, an energy density of about 1,935 joules/sq. inch (300 J/mm^2), the tip of a well-honed knife wielded by a strong adult has an energy density of 129,000 joules/sq. inch (20,000 J/mm^2)! This is why even though several kinds of bullet proof vests can block projectiles shot from a wide variety of firearms, they may be powerless to stop a simple knife thrust.

Of course, the effectiveness of bladed weapons is a direct function of how their blades are sharpened. Japanese master blacksmiths, for example, gained an almost perfect mastery of the art of making *katanas*, whose blades were so sharp and so hard that samurais could cut off an enemy's head with ease. The results obtained in Europe, however, were sometimes less spectacular, and the quality of blades occasionally left something to be desired. Thus, when Mary, Queen of Scots, was executed in 1587 for her alleged participation in a plot against Elizabeth I, the executioner's first ax blow made only a gash in the back of her skull. The second struck the nape of her neck but did not succeed in cutting

< Tip of a spear

through her neck completely, and it was only on the third stroke that her head was cut off. This was not an isolated case. In fact, the distressing spectacles of botched decapitations were what led Dr. Joseph Ignace Guillotin, to propose a more humane form of execution a few years later. This was to lead Antoine Louis and Tobias Schmidt to design the device that would bear the doctor's name: the guillotine.

Despite their strike force, bladed weapons have the major inconvenience of putting the aggressor close to the target, who can then in turn, strike a deadly blow. The possibility of killing from a distance was therefore a preferred strategy very early on in weapon development, as shown by the successive appearance of the spear-thrower or atlatl (20,000 years ago), the bow (12,000 years ago) and the crossbow (7,000 years ago). These devices all generated considerable force and were able to kill prey or an adversary rapidly and from a safe distance. The concept of killing from a distance reached its peak in the 20th century, with the invention of drones, airplanes without pilots that are capable of killing enemies while being controlled by human beings who may be thousands of miles away.

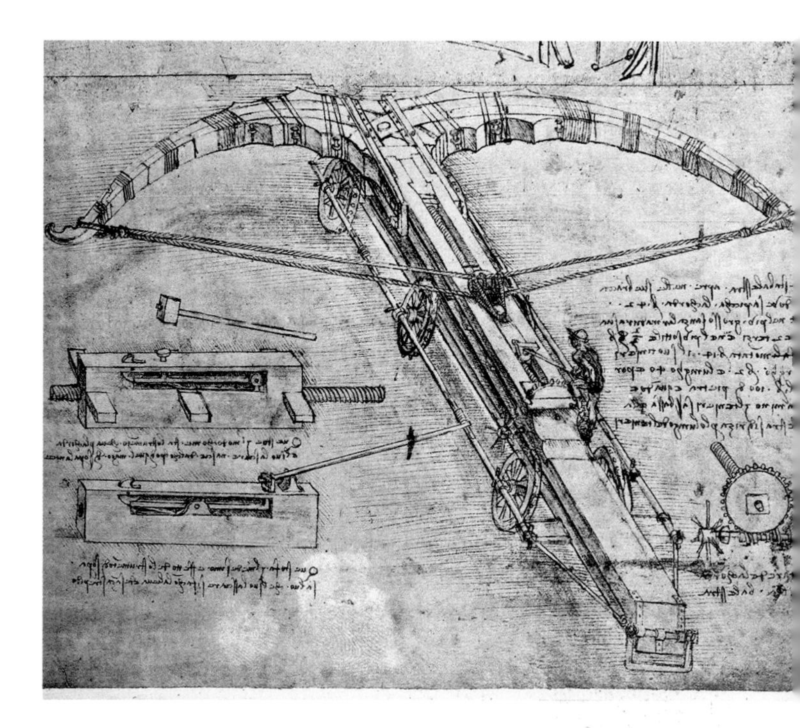

^ Two 16th-century Japanese *katanas*.

^ A study for a gigantic crossbow by Leonardo da Vinci.

A killing frenzy

The violence rife in the vast majority of societies, both primitive and modern, is seen in the innumerable massacres, human sacrifices, bloodthirsty tortures and bloody battles that have occurred throughout history. Whereas it is often believed that war is a modern invention of so-called civilized societies, a number of archaeological discoveries show, on the contrary, that many primitive peoples devoted a good portion of their time to waging war against neighboring peoples. For example, excavations at the ancient Nubian site of Gebel Sahaba, a cemetery that is roughly 14,000 years old, show that 40% of the skeletons of men, women and children have stone projectiles lodged in the bones, a sign of a violent death.

While we can imagine using violence to obtain food and even to fight a war, one of the most mysterious and horrible aspects of the history of civilizations is the enduring tendency of human beings to impose suffering on their fellow humans through torture. Stretching, quartering, crucifixion, impalement, flaying alive and other tortures have been used from time immemorial as punishment for serious crimes, to obtain criminal evidence (a confession or the denunciation of possible accomplices) or, even more horrible, to see someone suffer for the sake of pure sadism (see box on pp. 183–184). However, the power of torture to make someone tell the truth is greatly overrated, since suffering very often only serves to incite the victim to fabricate the response the torturers are looking for, without concern for its

^ A B-2 Spirit bomber accompanied by two F-117A Nighthawks.

Horrible deaths

Although by definition all forms of torture cause appalling suffering, three of them—quartering, crucifixion and impalement—stand out for their extreme cruelty. The torture of François Damiens, condemned to be quartered for having tried to assassinate King Louis XV, is described in this way in the edict announcing his execution:

> And on a scaffold to be built there, attached by his breasts, arms, thighs and the fleshy part of his legs, his right hand, holding the knife with which he attempted the said parricide, burned with fire and sulphur; and on the places where he is attached, molten lead, boiling oil, burning pitch, a mixture of melted wax and sulphur shall be poured; and then his body drawn and dismembered by four horses, and his limbs and body consumed by fire, reduced to ashes, and his ashes scattered to the winds.

The torture described was in reality much more dreadful. Despite the suffering inflicted, the limbs refused to separate from the body even when pulled by six horses, and the executioner had to slash at the thighs and shoulders so Damiens's limbs could be torn off and the sentence carried out.

Probably originating in Persia around the seventh century BCE, crucifixion is another cruel method of execution, chosen explicitly to cause a slow, extremely painful and humiliating death. The torture victims were usually beaten, whipped and then attached with rope or nails to a gallows in the shape of a cross. The wounds caused by the flagellation as well as the destruction of tissues by nails must have caused dreadful pain, all the more so since, as the vital organs were not affected, the torture victim's death throes could last a very long time. The causes of death apparently depended on the circumstances and the victim's overall physical state, but it is likely that exhaustion and hypovo-

^ Engraving representing the bloodthirsty Vlad III, known as the Impaler.

lemic shock caused by dehydration and blood loss, as well as suffocation, were among them.

These punishments were often used by the Romans to suppress local rebellions, and the executions sometimes reached heights of horror difficult to imagine. Thus, when in 71 BCE Crassus crushed the slave rebellion led by Spartacus, he had 6,000 slaves crucified along the 125 miles (200 km) of the Appian Way (*Via Appia*), linking Rome to Capua. The crucifixion of Jesus of Nazareth remains, however, the execution that has had the greatest influence on world history, his death on the cross being interpreted by Christians as a sacrifice for the redemption of the sins of humanity.

Impalement is definitely the only torture even more horrible than quartering and crucifixion. This technique—used on a large scale by the prince of Wallachia, Vlad III Basarab (1431–1476), known as Vlad the Impaler, involved driving a wooden stake (the pale) into the victim's body and then planting it in the ground so that the weight of the body would cause the stake to penetrate gradually. The torture was especially horrible when the tip of the stake was rounded and inserted into the anus, from where it would slowly pierce through the body and eventually emerge through the chest, shoulders or mouth, sometimes several days later. The terror inspired by the prince and the Sultan Mohammed's reaction when he tried to invade Wallachia are understandable:

> He could not help but react in horror when he saw before him a forest of stakes; spread over half a league's distance were more than 20,000 Turks and Bulgars, some impaled and others crucified. In the middle of them, on a higher stake, Hamza-Pacha could still be seen, clothed in his magnificent crimson silk robes. Next to their mothers were children, in whose entrails birds were nesting. On witnessing this theatre of atrocities, the fierce Sultan cried out: "It is impossible to drive out of his own country a man who has been able to commit such fearsome acts and who has contrived to use his subjects and his power to such effect. However," he added, no doubt sorry to have revealed his secret thoughts in this way, "a man who has committed so many crimes is not at all worthy of respect." (Chalcondyles, *L'histoire de la décadence de l'Empire grec et establissement de celuy des Turcs* [The history of the decline of the Greek empire and the establishment of that of the Turks], 1577.

The son of Vlad II the Dragon (Vlad Dracul), Vlad the Impaler was also known as Draculea ("little dragon" in Romanian), the nickname behind the bloodthirsty character created by Bram Stoker in 1897.

> An accused man is tortured in front of members of the court to extract a confession (engraving from a painting by A. Steinhell), around 1450.

truthfulness. For example, in the Middle Ages, a confession was considered to be incontestable proof of the accused's guilt. Consequently, several "questioning" techniques were perfected during this period in order to extract a "true" confession, whether through water torture, where the victim was forced to swallow several pints of water, or by crushing limbs using wooden planks (called brodequins), to name just two. The value of these "confessions" was more than dubious, since the accused would say anything at all to make the intolerable suffering stop. Most disturbing of all is the fact that these abuses were sometimes carried out under noble pretexts (at least from the authority's point of view!). Torture was occasionally even used by the Roman Catholic Church during the Inquisition.

Human beings are without doubt the animal that most fears death, but paradoxically we are also the one with the fewest scruples about causing our fellow humans to suffer and die—a killing frenzy that only got worse with the invention of explosives and firearms.

Setting off the powder keg

While the development of metallurgy led to the production of deadly bladed weapons, the discovery of gunpowder marked the real turning point in human beings' ability to use violence to get what they want. This black powder, composed of saltpeter (potassium nitrate), charcoal and sulfur, was, by all accounts, invented by the Chinese in the Tang dynasty (ninth century). The Chinese used it mainly for pyrotechnics (they liked to make rockets by stuffing bamboo stems with it). It did not take long, however, to discover the powder's deadly power, and it was used to propel ammunition at high speeds. The

^ Depiction of a Chinese celebration with fireworks.

only explosive known to humanity for over 500 years, gunpowder completely redefined the rules of war, both with regard to attack strategies and in the development of methods of defense against this new and powerful weapon (see box at right).

How can a simple powder, totally inert in its natural state, produce such an explosion when heated by a flame? Under normal conditions, the combustion of a substance depends on the availability of oxygen in the air. This is why fire is fanned by wind or, conversely, completely smothered if the supply of air is blocked. The ingeniousness of explosives lies in the fact that they contain both combustible substances and oxidizers, which are substances that can provide the oxygen needed for combustion even in the absence of air. In the case of gunpowder, oxygen comes from the saltpeter (KNO_3), which, in the presence of a heat source, causes the carbon atoms in the charcoal and sulfur to oxidize and form carbon dioxide and nitrogen:

$$10KNO_3 + 8C + 3S \rightarrow 2K_2CO_3 + 3K_2SO_4 + 6CO_2 + 5N_2$$

The explosion caused by gunpowder thus results from its ability to release, in a very short time, a considerable quantity of chemical energy as heat, thereby putting the gases produced in the reaction under high pressure and raising them to a high temperature. The heat generated, combined with the expansion of the gases, creates an explosive wave that travels at 1,310 to 2,625

Tanegashima

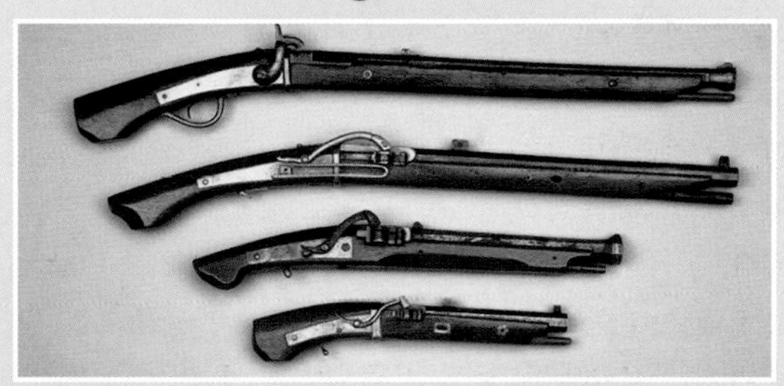

Firearms played a decisive role in ending the Sengoku period (known as the Warring States period), and the civil war that ravaged Japan from the 15th to the 18th centuries. In 1543, a Chinese ship carrying Portuguese passengers was wrecked on Tanegashima Island, south of Japan, with matchlock muskets on board. These weapons were very popular at that time but completely foreign to the Japanese, whose *bushis* (samurai) were still exclusively using bladed weapons (sabers, bows, lances). During the memorable battle of Nagashino on June 29, 1575, Oda Nobunaga and his ally, the future Shogun Tokugawa Ieyasu, used 3,000 muskets in three successive waves to completely destroy the cavalry of the powerful Takeda clan and win a decisive battle in Japan's reunification process. As told in the film *Kagemusha* by Akira Kurosawa (1980), the use of these *tanegashima* in the battle of Nagashino is an eloquent example of the power of firearms and, especially, of their influence on the course of history.

feet/second (400 to 800 m/s). When this explosion occurs in a confined space (e.g., a cannon or cartridge), the energy released is more than sufficient to eject a projectile at high speed. Although gunpowder has long since been replaced by more effective explosives based on nitrocellulose and nitro-glycerine—it is now used only in fireworks—the vast majority of homemade explosive devices used today still work on the same principle. For example, a molecule related to saltpeter, ammonium nitrate (NH_4NO_3), can be used to make explosive devices with frightening destructive power when mixed in appropriate quantities with a fuel. This molecule is found in a number of fertilizers and is thus an easily obtainable oxidizer, especially in Afghanistan, where it is used by the Taliban to make deadly bombs for use against soldiers in the NATO forces. In the United States, ammonium nitrate is infamous for its deadly role in the destruction of the Alfred P. Murrah Building in Oklahoma City by Timothy McVeigh in 1995, an attack that took the lives of 168 people, of whom 19 were children at a daycare center in the building.

Strike force

Since the main role of a weapon is to fire, as fast as possible, an object designed to kill (or at least neutralize) prey or an enemy, the discovery and mastery of explosive mixtures marked a great leap forward in the history of weapons, considerably increasing the speed and distance a projectile can cover (see Figure 1). For example, whereas the best archer will have difficulty hitting a target more

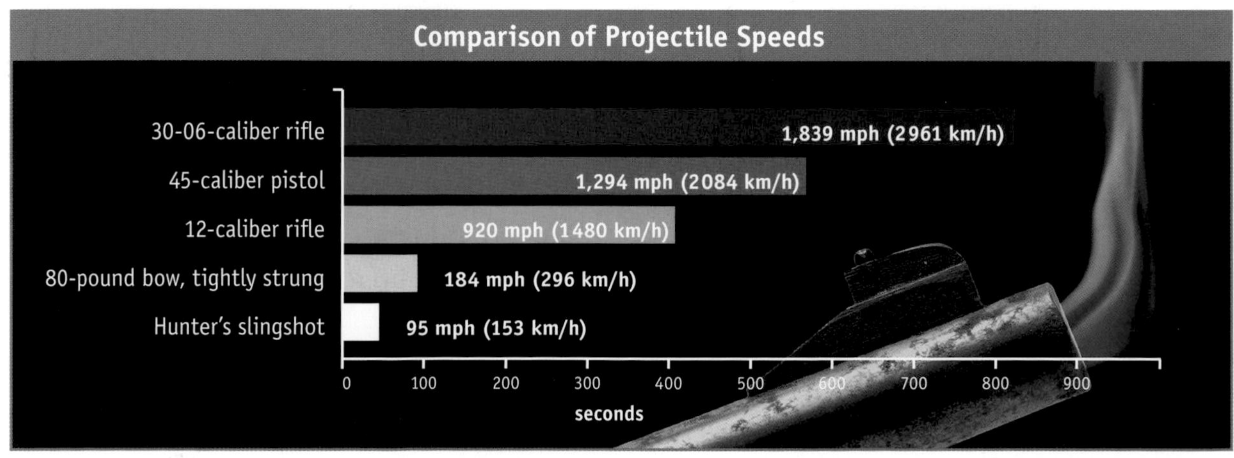

Figure 1

than 165 feet (50 m) away because of the earth's gravity, which inevitably pulls the arrow toward the ground, particularly talented snipers among the allied forces in Afghanistan have succeeded in hitting targets 1½ miles (2.4 km) away!

Modern firearm projectiles are made up of three main parts, the cap, the explosive charge and the shell itself, which are sealed inside a cartridge. When the trigger is pulled, a spring mechanism causes a small explosive charge in the cap to light, which literally "lights the powder keg" in the nearby fuel. At first this burns quite slowly (so that the weapon does not explode in the shooter's hands). The combustion slowly gathers speed and eventually creates a large amount of gas that shoots the projectile at very high speed toward the mouth of the gun. The explosion caused by the ejection of the bullet takes place in the confined space of the barrel of the weapon; thus, when the bullet is shot out of the gun, the pressure of the explosion is suddenly released, causing the bang typical of firearms.

From the point of view of physics, the energy associated with motion, also called kinetic energy, is expressed in the equation $E = \frac{1}{2} mv^2$, where m represents the mass of an object and v its velocity. According to this equation, it is possible to double the energy of a projectile by doubling its mass. However, if the velocity is doubled, its energy is quadrupled! Since the speed of a bullet is directly related to the quantity (and effectiveness) of the explosive charge in the projectile, bullets that are designed to travel long distances or to penetrate deeply into a very large animal (such as big game) have a larger volume that those used to hit closer targets.

However, increasing the impact force of weapons means not only increasing the speed of the projectiles but also the frequency with which ammunition is fired at the target. This is why Richard Gatling's invention of the machine gun in 1861 was a turning point in the history of firearms, marking the first stage in a frenetic race that resulted in the invention of many types of submachine guns, assault rifles and other automatic or semi-automatic weapons that can fire several thousand bullets in sustained bursts, making any movement by the adversary impossible. In their ability to shower an area with deadly projectiles, these weapons, just like grenades, bombs, land mines and other explosive devices, have completely redefined the ways of making war by making it possible to kill an adversary almost blindly, without the need for much precision.

Killing with bullets

Currently, the number of light weapons in the world is estimated at 688 million, 59% of which are owned by civilians, 38% by armies, 3% by police forces and 1% by illegal military groups. On the whole, it is estimated that firearms are annually responsible for 300,000 deaths in armed conflict and 200,000 deaths in civil society.

Mortality caused by firearms depends on four main factors: (1) a sufficiently deep penetration of the projectile into the body for it to reach vital organs, (2) the destruction of these organs by the formation of a cavity the diameter of the projectile (permanent cavity), (3) the formation of a temporary cavity, caused by the transfer of the projectile's kinetic energy on its way through the body and (4) whether there are bullet fragments or shattered bones that damage the organs (only in the case of high-speed projectiles; see Figure 2). Of all these factors, the location on the body where the bullet entered and the depth of penetration (primary cavity) are the most important parameters in determining the seriousness of the wound. Temporary cavities, however, play only a minor role in fatal wounds, as most of the tissues in the human body are elastic enough to absorb this shock and then return to their initial position without much damage. Nonetheless, some less elastic tissues, like the liver and spleen, or very fragile organs like the brain, can be damaged by temporary cavities.

Contrary to what one might think, it is very difficult to kill a person on the spot with a firearm. For example, one of the most persistent myths about these weapons, sustained by an unbelievable number of dramatic scenes on television and in the movies, is their allegedly powerful stopping power, as though it were enough to shoot at someone for him to be literally thrown backward the minute the bullet hits him. This kind of reaction makes no sense at all, as it is estimated that the amount of energy that hits the victim of a firearm is roughly equivalent to that of a baseball, a shock that is certainly very unpleasant but not enough to send a 150-pound (70 kg) human being reeling several feet backward. However, experience does show that many people fall to the ground when they are hit by a bullet, a reac-

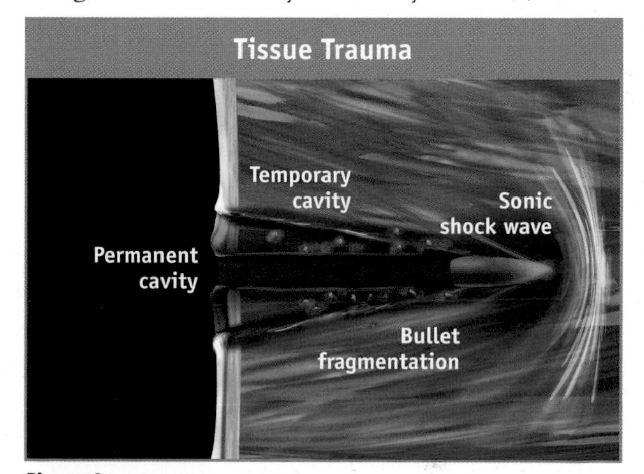

Figure 2

tion that is much more psychological than physical, the expression of an instinctive withdrawal behavior in the face of danger. This psychological element is clearly demonstrated by the absence of reactions in intoxicated individuals (alcohol, drugs) when they experience the shock of being hit by a bullet; this insensitivity makes them that much more dangerous in a violent context.

It is nearly impossible to instantly kill someone with a firearm, largely because it is difficult to precisely hit a part of the body whose destruction would cause vital functions to instantly stop. Although it looks dramatic, a bullet entering the body usually destroys only about 2 ounces (50 g) of tissue, a negligible quantity for a 150-pound (70 kg) individual if the part of the body hit is not critical to immediate survival. In reality, to kill someone immediately with a firearm, you have to hit the nervous system (brain or spinal cord) or cause a hemorrhage by hitting the main blood vessels or the heart. Even in the latter case, death is not immediate. In fact, even when circulation stops completely, there is still enough oxygen in the brain for voluntary movements to occur for up to 15 seconds after the heart is destroyed by a bullet, a very dangerous situation for police officers confronting a criminal, especially when the latter's psychological state (survival instinct, level of intoxication, aggressiveness) causes a powerful physical reaction that can overcome the bullet's effects. Obviously, the longer-term presence of a foreign body in the organism causes many complications that can ultimately lead to the victim's death. The length of survival depends on the importance of the affected organ, the speed of blood loss and, in the longer term, the contamination of the wounds by pathogenic microorganisms.

Immediate, violent deaths attributed to firearms are caused, most of the time, by a direct attack on the nervous system (a bullet that hits the brain or spinal cord) or by hemorrhaging caused by the destruction of major blood vessels. These effects are closely linked to the kind of projectile, its speed and its ability to penetrate the organism. Although the likelihood of being

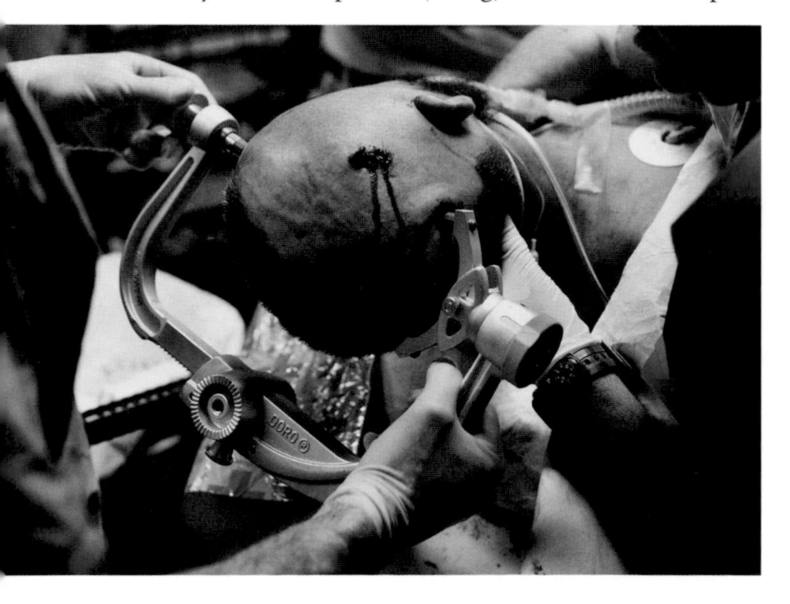

^ Preparing for a stereotacic operation to remove a bullet lodged in the cranium.

killed by a bladed weapon or a firearm is, for most of us, very low nowadays, there are still common objects in our daily lives that have a destructive power similar, in terms of physics, to that of firearm projectiles. This is particularly true of motorized vehicles.

Fatal collisions

Every year, motor vehicle accidents are responsible for the deaths of 1.2 million people worldwide, in addition to 140,000 injured and 15,000 people who will be handicapped for life. It is surprising to realize the degree to which our industrialized societies, where cars are everywhere, have accepted these accidents as inevitable, whereas the majority of them could be avoided by changing behavior behind the wheel. Alcohol is obviously a crucial factor in increased accident risk, but excess speed, often encouraged by the publicity of car companies themselves, is another significant factor. If we add to that a simple lack of attention, along with activities incompatible with driving a car, such as phone calls, text messaging, reading a newspaper, eating or putting on makeup, we have a perfect mix of risky behind-the-wheel behaviors.

The often fatal outcome of accidents is due to the powerful impact forces exerted on the body

during a collision. In practice, the seriousness of the traumas connected with accidents is determined by the fundamental laws of physics, as set out by Isaac Newton more than three centuries ago (see box on p. 194).

According to Newton's first law, at the moment a head-on collision occurs, each passenger keeps moving forward at the same speed as the car, which, without seatbelts, results in the ejection of the body, pure and simple. Buckling up really is worth it: during a collision at 50 mph (80 km/h), every passenger hits the steering wheel or the windshield in just 0.7

seconds! The same principle applies to our internal organs: when the body stops suddenly, each organ keeps on moving. This motion has the effect of considerably increasing the organ's "apparent weight" following a sudden deceleration (see Figure 3). A deceleration of less than 12 mph (20 km/h) is not generally enough to cause serious injury. However, as soon as the deceleration reaches about 22 mph (36 km/h), serious injuries can occur and the risk of serious trauma increases enormously at higher speeds.

In addition to the effects of sudden deceleration, the kind of obstacle struck by the vehicle in motion obviously has a major effect on the seriousness of the collision. Hitting a snow bank does not cause the same damage as hitting a concrete wall at full speed! The laws of physics enable us to calculate that a car moving at 31 mph (50 km/h) will create an impact force of roughly 50 tons in a brutal collision.

Newton's laws of mechanics

- **Principle of inertia:** An object in motion continues to travel in a straight line at a constant velocity when no outside force is applied.
- **Principle of dynamics:** The total force acting on an object is equal to the mass of the object multiplied by its acceleration (F = ma). Thus, the greater the deceleration (that is, the faster the loss of speed), the greater the force applied to the body and the higher the risk of injury in case of an accident.
- **Principle of action and reaction:** A body exerting a force on another body is subject in return to a force of the same magnitude but in the opposite direction. In other words, when a collision occurs, the amount of force that the body suddenly applies to an obstacle is equal to the amount of force that the obstacle applies to the body.

Impact of Speed

Organ weight in pounds (kg)	Apparent weight in pounds (kg) based on different collision speeds		
	22 mph (36 km/h)	45 mph (72 km/h)	67 mph (108 km/h)
Spleen 0.5 (0.25)	5.5 (2.5)	22 (10)	55 (25)
Heart 0.7 (0.35)	7.7 (3.5)	30 (14)	70 (31.5)
Brain 3.3 (1.5)	33 (15)	132 (60)	298 (135)
Liver 4 (1.8)	40 (18)	160 (72)	357 (162)
Blood 11 (5)	110 (50)	440 (200)	992 (450)
Whole body 155 (70)	1,543 (700)	6,173 (2,800)	13,890 (6,300)

Figure 3 Source: J. Albanèse, *Le Polytraumatisé* (2002)

Head injury is one of the most tragic consequences of motor vehicle accidents and is the primary cause of death in Canadians under 45. Despite being well protected by the cranium and several layers of resistant tissue, the brain nonetheless remains a "mobile" organ, likely to change position suddenly as a result of an impact and to bounce hard against the inside of the skull (see Figure 4). The seriousness of the ensuing trauma normally depends on the strength of the impact. In the case of a concussion, for example, the shock can cause a loss of consciousness that lasts from a few seconds to a few minutes, and the person may feel dizzy and lose vision or balance. While very frequent in contact sports, these concussions may sometimes need an extended period of recovery (and even put an end to a sports career). Cerebral contusions, on the other hand, are even more serious, as they often lead to shock and tissue damage that causes bleeding and a buildup of liquid (edema) likely to damage nerve cells (see Chapter 5). This situation is particularly dangerous when the blood pools and clots, forming a hematoma. In the latter case, the injured person often begins to feel unwell several hours or even several days after the shock (dreadful headaches, loss of balance, unusual behavior) and subsequently falls into a fatal coma. Finally, in some cases, the impact causes a crack in the cranium, and this fracture can also result in an accumulation of blood in the brain as well as direct damage to nervous tissue.

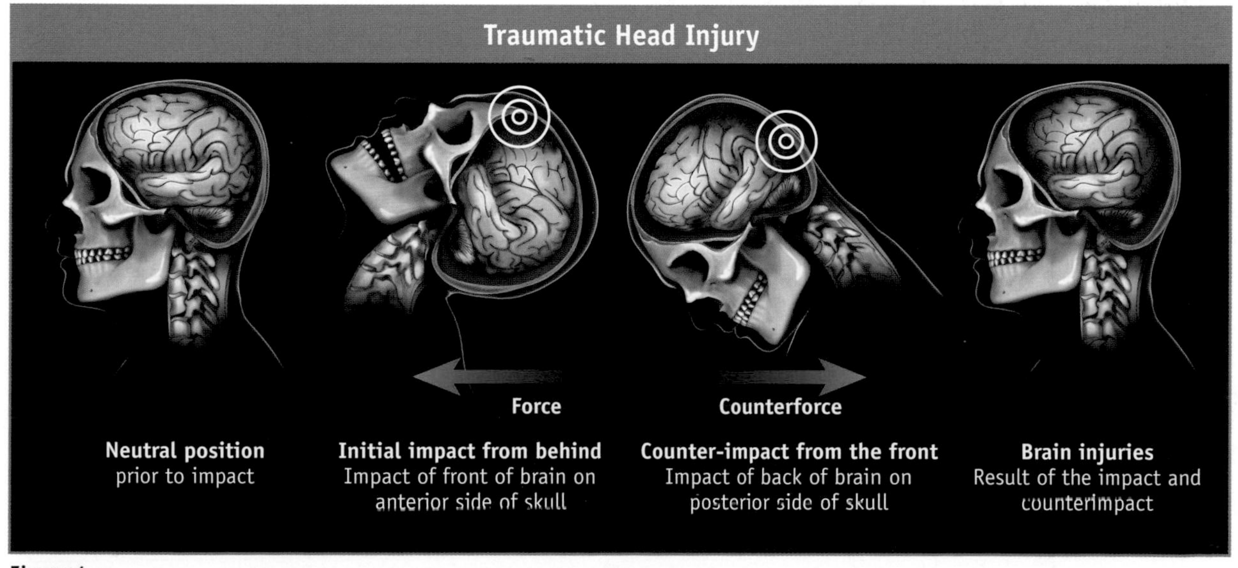

Figure 4

Fractures of the cervical vertebrae, also called "whiplash" (referring to an abrupt snapping motion like the crack of a whip), are also frequently seen in victims of motor vehicle accidents, especially rear-end collisions. This kind of impact causes hyperextension of the neck followed by hyperflexion. This violent back and forth movement can cause vertebral fracture. When the second cervical vertebra (the axis) is affected, the spinal cord can snap at the innervation center for the diaphragm, which controls autonomous respiration, and death quickly follows. If these vertebrae are spared, the victim can survive but often at the cost of being a permanent quadriplegic.

Fractures of the pelvis and the diaphyseal femur are also very often seen in victims of motor vehicle accidents. These fractures, which usually occur when the passenger strikes the dashboard at the moment of impact, cause serious bleeding and may lead to death from hypovolemic shock if blood loss is not brought under control quite quickly.

As with death caused by firearms, hemorrhagic shocks (especially those caused by thoracic trauma) and head injuries are the main causes of death from motor vehicle accidents (as well as from all accidents). The car can therefore be considered, just like a bullet, to be an actual projectile with the power to kill, which sadly happens all too often. The technical skills of car manufacturers have made vehicles as fast as they are efficient, but the comfort and soundproofing of the interior and their smooth handling keep us from gauging the true speed of the vehicle and the strength of the physical laws at play. What's more, there are more and more things to distract us while driving: cell phones, in-car computers , GPS systems and the complexity of the car's controls. We wind up underestimating the energy that a vehicle weighing several tons traveling at high speed can produce; we forget that cars can be extremely destructive in comparison with the fragile tissues that make up the human body.

^ David Booth, of the Florida Panthers, remains stretched out on the ice after a violent blow from Mike Richard, of the Philadelphia Flyers.

336

中村諌助匡辰
中村諌助匡
應需
一筆菴誌
一勇斎國芳画

Chapter 9

Unusual and Dramatic Deaths

Falling doesn't matter
Before everything else, before all the others
It is peculiar to the cherry tree's flowers
To fall with nobility
On a stormy night.
Yukio Mishima (1925–1970)

Honor or death

This poem—composed in 1970 by the great Japanese writer shortly before he took his life by *seppuku*, commonly called hara-kiri—shows how differently the Japanese and Westerners view death. Whereas we choose to stress the tragic and even unjust nature of death, the traditional Japanese view emphasizes instead the transient nature of life. Death is considered as natural an occurrence as flowers falling from a tree.

Historically, this detachment with regard to death is well illustrated by *seppuku*, the ritual suicide performed by dishonored samurais following a breach of the warriors' code of honor (the Bushido). Bu-shi-do literally means "the path of the warrior," or the moral code warriors must observe, both on the battlefield and in their day-to-day life. This unwritten code, refined and passed down from one samurai to another, is founded on the basic principles of Buddhism—respect for others, love of nature, confidence in fate, a calm acceptance of the inevitable nature of events—as well as on courage, a sense of justice and loyalty to one's master.

When one of these principles was not respected, the samurai preferred death to life. After a session of meditation and the writing of a final poem (*jisai*), the warrior, wearing a white kimono, wrapped the blade of his *wakizashi* (a short sword) in a piece of white cloth. He then inserted it sharply into the left side of his

< Artist Utagawa Kuniyoshi's depiction of samurai Nakamura Kansuke Tadatoki.

The 47 Ronins

One of the most famous Japanese stories is that of the 47 Ronins, which was immortalized by one of the great masters of woodblock printing, Utagawa Kuniyoshi (1797–1861). This epic, entitled *Chushingura* or *Treasury of Loyal Hearts*, illustrates the code of the warrior (the Bushido). The events took place in 1701, when Lord Asano Nagamori was forced to commit suicide by the shogun because of a conflict with Lord Kira. Asano's belongings were confiscated, and his samurai once again became ronins, or masterless. After preparing for a year, the 47 most loyal samurai of Lord Asano successfully avenged him by attacking Kira's castle and killing him. They then committed ritual suicide, having fulfilled their duty of loyalty to Asano. The memory of the 47 Ronins is still honored today, notably in a temple dedicated to them in the center of Tokyo. It depicts the sacrifice of the individual to a code of honor and represents one of the aspects of the Japanese soul.

abdomen in line with his navel, and then he made a long incision across to the other side. In some extreme cases, he made a second vertical incision from bottom to top, eviscerating himself and rupturing major blood vessels such as the mesenteric arteries and the aorta. The pain was no doubt dreadful, and the person committing suicide needed an assistant (a *kaishaku*), often a friend, to cut short his death throes by decapitating him using his *katana* (a long sword). It is difficult for Westerners to accept such a ritual, but in feudal Japan, *seppuku* was a mark of bravery, a way for a warrior dishonored by defeat, treachery or serious injury to put an end to his life and thus preserve his honor.

A noose around the neck

> And if any mischief follow, then thou shalt give life for life, eye for eye, tooth for tooth, hand for hand, foot for foot, burning for burning, wound for wound, stripe for stripe.
>
> Exodus 21:23–25. *King James Bible*

The establishment of a criminal justice system controlled by the power of the state was a key step in maintaining order in human societies. By giving the state a monopoly over violence, these systems led to a decrease in the numbers of personal vendettas and allowed for the establishment of an independent justice system, which was intended to ensure the common good

< Lithography of a samurai preparing for *seppuku*.
< Kaga-style armor from the Edo period.

of the population. Talion law, the subject of the above quote, is an early example of this kind of system, a form of justice that (fortunately) has become more sophisticated over time. However, no justice system is perfect, since its fairness is a direct function of the ideology of the ruling power. Sadly, the abuses perpetrated through state-controlled violence are too many to count —demonstrations of power whose main objective is too often to create a climate of terror in the very population that the state is supposed to protect. Even today, several totalitarian states use the execution of "criminals" who oppose the regime as a means of maintaining their control over the population.

For a long time, hanging was (and still is in certain totalitarian countries) the preferred method of executing people found guilty of crimes deemed sufficiently serious to warrant death. While this form of execution is effective, it can be cruel, depending on the procedure used. Indeed, contrary to what is often believed, in the majority of cases it is not the blocking of air entering the lungs that causes death, since strong pressure is needed to stop air from circulating in the airways, as the windpipe is well protected by rings of cartilage. In general, the tightening of a noose around the neck mainly affects the blood vessels that connect the brain to the rest of the body, and, depending on the pressure exerted, the compression of these vessels may sometimes not

be enough to cause a quick death. Several types of hanging have been practiced throughout history, the most common of which are:

Short-drop hanging The body of the condemned drops from a low height, and it is the person's weight that tightens the knot around the neck. The nearly universal hanging method for most of history, this method of execution is still employed today in certain countries, notably Iran. Hanging is the preferred means of suicide in most parts of the world, particularly in Eastern Europe, where it is used by 90% of those committing suicide.

Death caused by this kind of hanging is in every way similar to that caused by strangulation, except that the pressure on the neck comes from the weight of the body, whereas in the case of strangulation the pressure is exerted by hands (normally those of a murderer). It is mainly the unfortunate condemned person's jugular veins that are affected, which prevents blood in the head from flowing back to the heart. This blockage causes an accumulation of blood in the head, seen by the swelling and blue coloration of the face, as well as edema in the brain, ultimately causing a loss of consciousness. Unfortunately, this last stage may only happen after long and terrible death throes, in which the condemned only dies after jerking and twitching at the end of the rope. Since executions by hanging were often car-

^ Four criminals are hanged at a public execution in Shiraz, Iran, on September 5, 2007.

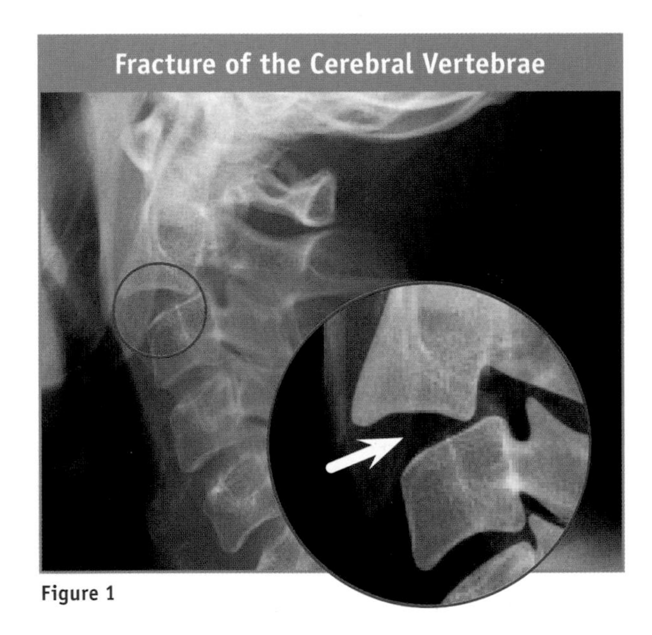

Figure 1

ried out in public as a way of setting an example, these "dances of the hanged" were a spectacle usually much enjoyed by curious onlookers. A variation of short-drop hanging is hanging by suspension, where the condemned is hoisted into the air (using a crane, for example). The factors causing death from this form of hanging are essentially the same as those for short-drop hanging (it is still the weight of the body that creates tension in the rope).

In contrast, in some cases, the noose compresses the carotid arteries and immediatley blocks the blood supply to the brain, which is rapidly followed by loss of consciousness. The compression of the carotids also puts pressure on the carotid body, a structure full of blood vessels located at the fork of the carotid artery, which can precisely detect the pressure of the blood flowing to the brain. Its sensors will interpret the force applied by the rope as a sign of elevated tension and immediately slow down the heartbeat. This, in extreme cases, triggers cardiac arrest. Between 6 and 15 seconds are needed for the compression of the carotids to cause loss of consciousness, with death occurring about 5 minutes later. A number of martial arts blows (*atemi*), accompanied by violent battle cries (*kiai*), target these vulnerable spots, long known by the Japanese masters of unarmed combat to be critical to life.

Long-drop hanging This form of hanging was invented for "humanitarian" reasons, with the purpose of causing instant death and thus avoiding the needless suffering of short-drop hanging. The condemned drops from a certain height through an open trapdoor in such a way that, at the end of the drop, the acceleration of the body is suddenly stopped by a knot in the rope, rupturing the spinal cord and causing rapid death. In practice, however, the length of the drop, generally equal to the height of the condemned person, was often too short to break the neck, and the victim died of strangulation. However, too long a rope could cause complete decapitation, a death no doubt rapid, but nevertheless horrible and traumatic for the executioners. In 1872, in order to avoid these extremes, an English executioner, William Marwood, devised a way to calculate the drop required based on the weight of the condemned. It provided a "calculated death," where the force applied on the neck would cause the dislocation of the second cervical vertebrae, a fracture known since as "the executioner's fracture." Today it is usually seen in the wake of certain car accidents (see Figure 1).

Deadly drugs

> Happiness is continuing to desire that which one possesses.
> St. Augustine (354–430)

One of the most bizarre phenomena resulting from an increase in cerebral capacity in mammals is the brain's talent for creating a totally false reality, a virtual world that usually appears in the form of more or less bizarre dreams during sleep. It is difficult to measure in any concrete way the influence of these dreams on the human psyche, but what is certain is that they indicate hyperactivity in the brain, almost as if its abilities were far greater than the physiological function it fulfils and the reality of the everyday world is not enough to satisfy it. It is often suggested that dreams are involved in sorting through the information accumulated by the brain, as well as in storing memories and selecting relevant data.

A good illustration of the propensity of the human brain to constantly try to push back the limits imposed on it by reality is the significant role that psychoactive substances have played throughout history. Whether for their ability to make us more alert (coca, tobacco, coffee), alter how we perceive the world (alcohol, cannabis, opium) or even take us to new levels of parallel consciousness through auditory and visual hallucinations (magic mushrooms, mescaline, iboga, ayahuasca), all cultures have, in one way or another, given particular status to these substances, for both medicinal and religious purposes.

Among all the psychoactive substances identified over the centuries, the latex sap produced by certain species of poppy has long been known to be one of the drugs most able to alter perception. Called the "plant of happiness" by the Sumerians (3000 BCE), the poppy was

^ Bottles of laudanum

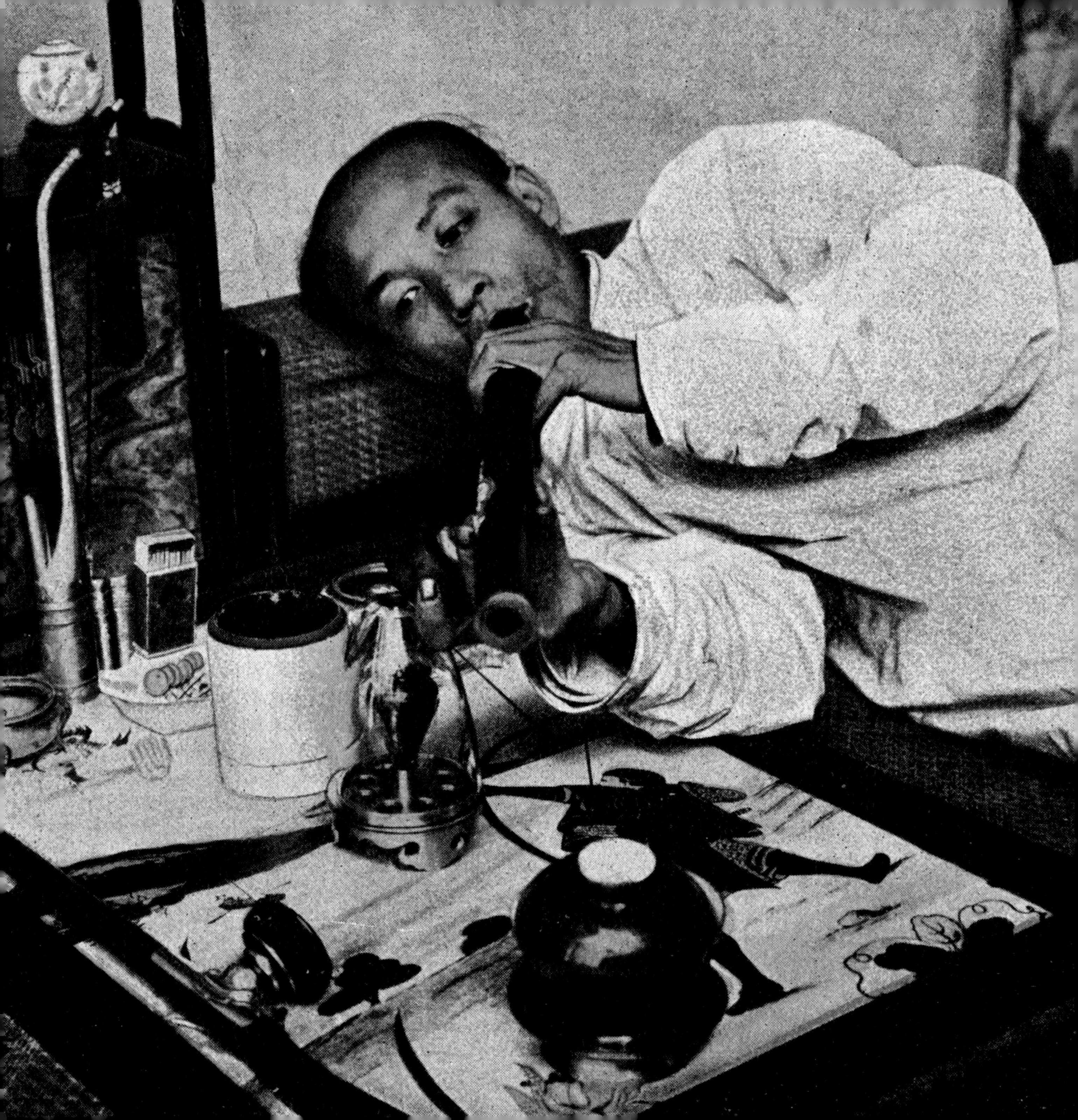

Trapped in the arms of Morpheus

Opium is extracted from the *Papaver somniferum*, a very beautiful variety of poppy cultivated for several thousands of years in Mesopotamia and southern Europe. When the plant reaches maturity, its head secretes a white latex sap containing about 60 complex alkaloids, morphine in particular. When these capsules, called *kodeion* by the Greeks (hence the name "codeine" for one of the poppy's alkaloids), begin to turn yellow, an incision is made to extract the *opios*. This sap takes on the consistency of a resin as it dries and may contain up to 16% morphine. By acting on the limbic system (the seat of emotions), this molecule suppresses pain and tension, causing a mild euphoria. Repeated consumption of opiates, however, produces a tolerance to its effects; this desensitization then requires increasingly higher doses to obtain a similar effect, thus creating a very powerful addiction. This addiction is particularly powerful in the case of heroin (diacetylmorphine), a molecule not found in nature that is four or five times stronger than morphine, owing to the greater ease with which it reaches the brain (between 15 and 30 seconds after being injected, 7 seconds when it is smoked). Synthesized from morphine by chemists working for the German pharmaceutical company Bayer, heroin was first marketed in 1898 as a substitute for morphine and a cure for coughing (tuberculosis and pneumonia were the main causes of death at the time). It received a warm reception from the medical community, and it was only several years later that people reailzed heroin was rapidly metabolized into morphine and, as a result, created a strong dependency. Some users were prepared to do anything to find the money they needed to buy heroin-based drugs, even going through garbage bins in search of any object that could be sold (hence the label "junkie" for heroin addicts). No other molecule is more directly associated with the ravages caused by drug addiction than heroin, a drug that unfortunately remains accessible to nearly 20 million users worldwide, thanks to *Papaver somniferum's* intensive cultivation in the Golden Triangle (Laos, Burma, Thailand) and the Golden Crescent (Iran, Afghanistan, Pakistan).

< Postcard of a Chinese man smoking opium (around 1900).

mentioned in the *Codex Ebers*, an Egyptian medical papyrus dating back to 1500 BCE, as a remedy (undoubtedly quite effective!) for "keeping babies from crying." The Greeks were fascinated by the poppy's psychotropic effects, both for soothing pain and for reaching a state conducive to communicating with the gods. What's more, in Greek mythology, Hypnos (god of sleep) and his son Morpheus (god of dreams) usually appear holding poppy flowers to give humans peaceful rest and pleasant dreams.

The sap of the poppy, better known as opium, is a complex mixture of several plant metabolites (sugars, lipids, proteins, gums, wax) and some 60 alkaloids, of which the best known are morphine (from 10% to 15%), codeine (1% to 3%), noscapine (4% to 8%), papaverine (1% to 3%) and thebaine (1% to 2%). All these molecules (with the exception of thebaine) have outstanding analgesic properties and are able to suppress pain. Morphine is still an indispensable drug for easing the suffering brought on by very serious diseases, particularly for cancer patients in their terminal phase.

Morphine's analgesic effect is due to its ability to imitate the action of endorphins, a class of neurotransmitters produced by the brain in response to pain. These endorphins (proenkephalin, prodynorphin, proopiomelanocortin) bind with a family of receptors located mainly in the limbic system (the seat of emotions), causing states of happiness, relaxation, courage and pain tolerance (in addition to pain, these neurotransmitters are also released in response to exercise and intense sensations such as orgasm). By binding to these receptors, morphine does not, therefore, block the mechanisms responsible for pain as such, but rather the subjective perception of the pain. It is very common for a person taking morphine to be conscious of the existence of pain while being completely unaffected by it.

The important role of endorphins in our "emotional health" means that activating the pleasure pathways using opiates like morphine is pleasing to the brain. It should not, therefore, be surprising that the use of opium has only increased with time! In the 16th century, the alchemist Paracelsus (1493–1541) discovered that extracting opium using alcohol (in his case brandy) produced a drug so effective that he named it laudanum, from the Latin *laudare*, meaning "praise." Very popular until the end of the 19th century as a universal medicine for treating an impressive array of disorders, from the common cold to heart diseases, laudanum was even used by exhausted parents as an infallible way to calm their children down.

Unfortunately, repeated consumption of opiates leads to serious addiction (see box on p. 207). Opium's extremely widespread availability, made possible by the colonial empires, especially the British, laid the groundwork for what would become a serious societal problem in the 20th century—narcotic drug addiction. "Oh just, subtle and powerful opium... you hold the keys to paradise!" wrote Baudelaire (in *Les Paradis artificiels* [The artificial paradises], 1860), echoing the widespread popularity of the drug at the time. This is no doubt true in the very short term, but the following decades would cruelly show the dangers associated with the search for this kind of false paradise, not only with opium and its derivatives, but also a range of synthetic mole cules that affect the nervous-system, like benzodiazepines (Ativan), barbiturates and other nervous -system depressants.

Respiratory depression

The well-being and the relaxing effect narcotics such as opium, alcohol, benzodiazepines and barbiturates provide are, however, accompanied by an interaction of these molecules with the neurons of the brain stem involved in controlling respiration. When too large a dose of one or another of these drugs is taken, the depressant effect becomes too powerful and can completely stop the nerve impulse from reaching the lungs, causing death from respiratory failure. This

^ Engraving, based on a painting by Michelangelo, showing Phaeton, the son of Helios, falling to earth after being struck by a thunderbolt thrown by Zeus.

danger is particularly high among users of strong opiates such as heroin: death rates among regular users of this drug are up to 20 times higher than among non-users, particularly due to overdose. However, contrary to popular belief, these deaths are rarely due only to an overly high quantity of the drug in the blood, but rather to an interaction with other depressants.

Lightning

> One is never struck if one has seen the lightning or heard the thunder.
> Pliny the Elder, *Natural History*, II c. 77–79

Lightning is one of the most common and spectacular violent natural phenomena: every second, 2,000 thunderstorms strike the planet. There are on average 45 lightning bolts per second, amounting to 1.5 billion per year. In all civilizations, lightning was considered to be the sign of a gods' anger over the actions of human beings. Whether it was Zeus for the Greeks, Adad for the Babylonians, Indra in India or Raijin in Japan (to name only a few), divinities had the capacity to cause thunder and lightning.

Because of this lack of understanding of the phenomena, lightning has for thousands of years taken humans completely by surprise, sometimes leading them to adopt extremely dangerous behaviors. Thus, it was long believed that ringing church bells could stop thunderstorms and make lightning strike further away, a little like a prayer sent to God asking him to spare humans. The effectiveness of this strategy left a lot to be desired however, as lightning very often struck the church steeple itself! A census taken in 18th-century Germany shows that in barely 30 years lightning had struck 386 church towers and 121 people had been killed or seriously injured by lightning while ringing church bells during a thunderstorm. We now know that these poor bell ringers were in the worst possible place, as a steeple is the ideal point of contact between the ground and the electricity in the clouds (see box on p. 213).

Lightning deaths

Since the human body contains a significant quantity of electric circuits, composed of nerves and essential to maintaining vital functions, it is very dangerous to be struck by lightning, as the high intensity of the electric discharge transmitted to the body can disturb the normal transmission of nerve impulses. This is why death caused by lightning, as with death from all forms of electrocution, usually results from the failure of the functions that are the most dependent on nerve signals, in particular the cardio-respiratory functions. However, in addition to its effects on organs, accidents caused by lightning have a wide

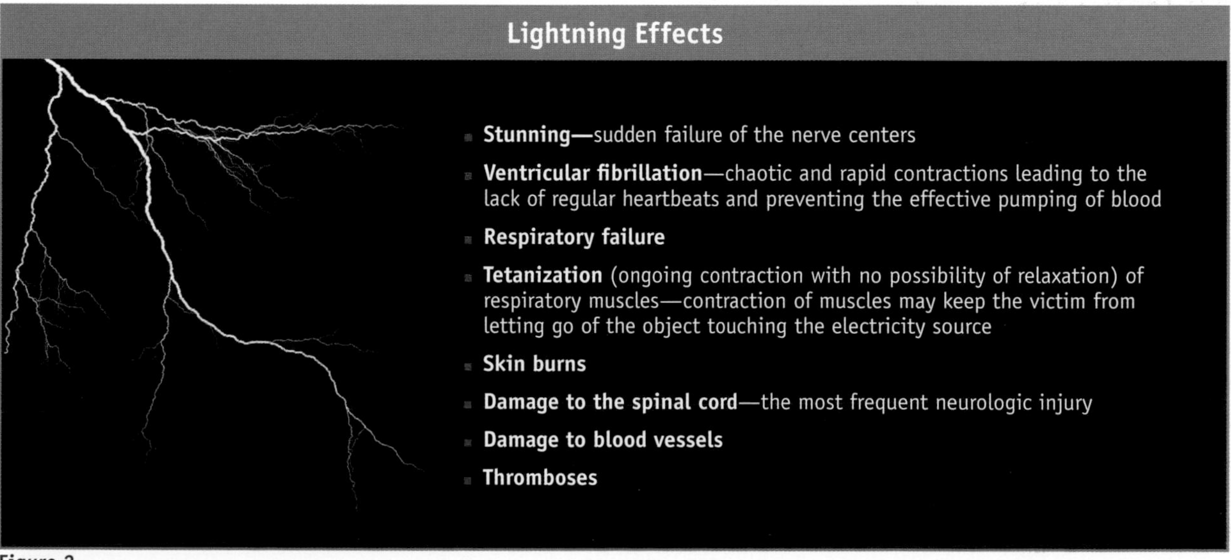

Figure 2

How is lightning produced?

The clouds responsible for thunderstorms (generally cumulonimbus clouds several miles thick) acquire an enormous electric charge as they form. The summit of the cloud mass bears a positive charge while its base (the side toward the ground) bears a negative charge. The two regions with opposite charges thus create an electric field that may create a current inside the cloud mass itself. This then produces lightning bolts in the sky (three-quarters of the lightning flashes in a thunderstorm are of this type). What is more dangerous for us, however, is that the presence of negatively charged particles on the underside of these clouds produces an accumulation of positive charges on the ground (since opposite charges attract each other). Although the atmosphere is not a good conductor of electricity, its insulating power cannot stop the attraction between these charges when the intensity of the cloud's electric field becomes very strong. A few invisible negative charges (called "descending tracers") head toward the ground, causing the positive charges on the ground to move. These charges tend to accumulate in elevated locations (like a church tower, treetop or a person standing upright), and when the descending tracer is near enough, an "ascending tracer" is created at the top of the object on the ground's surface. At this stage, there really is "electricity in the air," and the impending meeting between the two tracers may be indicated by a bluish glow and crackling sounds atop the elevated structures, such as the masts of a boat (called St. Elmo's fire), or, in the case of a person, by the hair standing up on the head. When the two tracers meet, a conductor bridge is created between the ground and the sky, and an intense electric current flows through the channel thus formed: a gigantic short-circuit with an intensity of 10,000 to 25,000 amperes at a tension of 10 to 100 million volts shoots upward (from the ground to the cloud) at a speed of over 60,000 miles/second (100,000 km/s). The violence of this electrical flow causes the air to heat up dramatically (up to 54,000°F/30,000°C!) and produces a typical lightning "flash." The increased air temperature from the enormous burst of heat also creates a sound wave that spreads out from the point of impact, causing the clap of thunder associated with lightning. The thunderclap is only heard after the lightning, as the speed of sound is much slower than that of light. To calculate the distance in miles between yourself and the spot where lightning struck, all you have to do is count the seconds between the appearance of the lightning and the moment when you first hear the thunder and divide this number by five (three, if measuring in kilometers). Since the bolt of lightning starts on the ground, the thunder also starts on the ground—hence the deafening noise heard when lightning strikes nearby.

variety of sometimes very dramatic characteristics (see Figure 2).

Although it usually causes the death of only one person at a time, lightning remains one of the most deadly natural phenomena, ahead of even tornados and hurricanes in terms of the annual number of victims. Roughly 10% of people struck by lightning die from it, and 70% suffer serious long-term consequences, particularly memory loss and major personality changes. Some unlucky people have been struck by lightning several times in the course of their lives. Roy Sullivan—a forest ranger from Virginia nicknamed the "human lightning rod" after having been struck no fewer than seven times between 1942 and 1977—holds the absolute record. He survived each of these strikes but lost a toenail and his eyebrows and suffered a variety of wounds to his arms, legs, chest and stomach.

A direct lightning strike This happens when a person in contact with the ground is located in the path of the flash, usually out in the open. The current circulates in this case between the highest part (the head or an object held over it, like an umbrella) and the ground, passing through the legs. When a charge travels through the body, its

^ Sliding panel depicting a god of thunder by Suzuki Kiitsu.

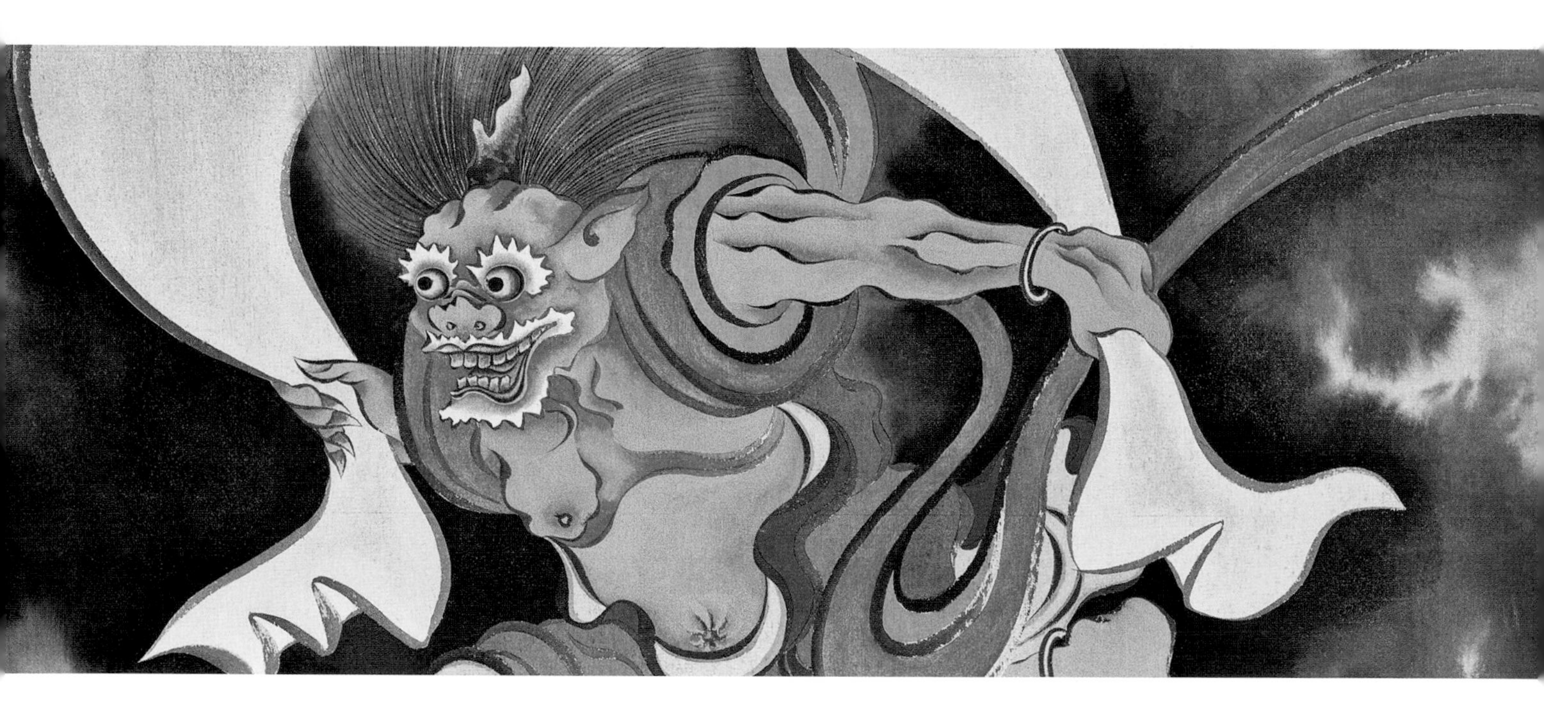

intensity peaks at around 1,000 amperes after just a few microseconds. The difference in potential between the head and feet is 300,000 volts. Since, fortunately, the human body's resistance is relatively high, the majority of the current takes a path of less resistance (a bypassing arc) and flows to the outside, on the surface of the body. The intense heat associated with such high voltage may nonetheless cause sweat to evaporate, clothing to vanish into thin air (including shoes or boots) and burns from the melting of metallic objects (such as a belt buckle) in contact with skin. In spite of all of this, thanks to this bypass, the current passing through the body is 3 amperes on average and lasts 10 to 20 milliseconds. This intensity has various effects, but none, however, is life threatening. The flow of most of the current through a bypassing arc and the usually brief duration of the current's passage through the body both have a protective effect, limiting the risk of ventricular fibrillation and internal electrothermal burning.

Nonetheless, the intensity of the current from the flash passing through the body often has serious pathological consequences that may cause death.

^ Sliding panel depicting a god of wind by Suzuki Kiitsu.

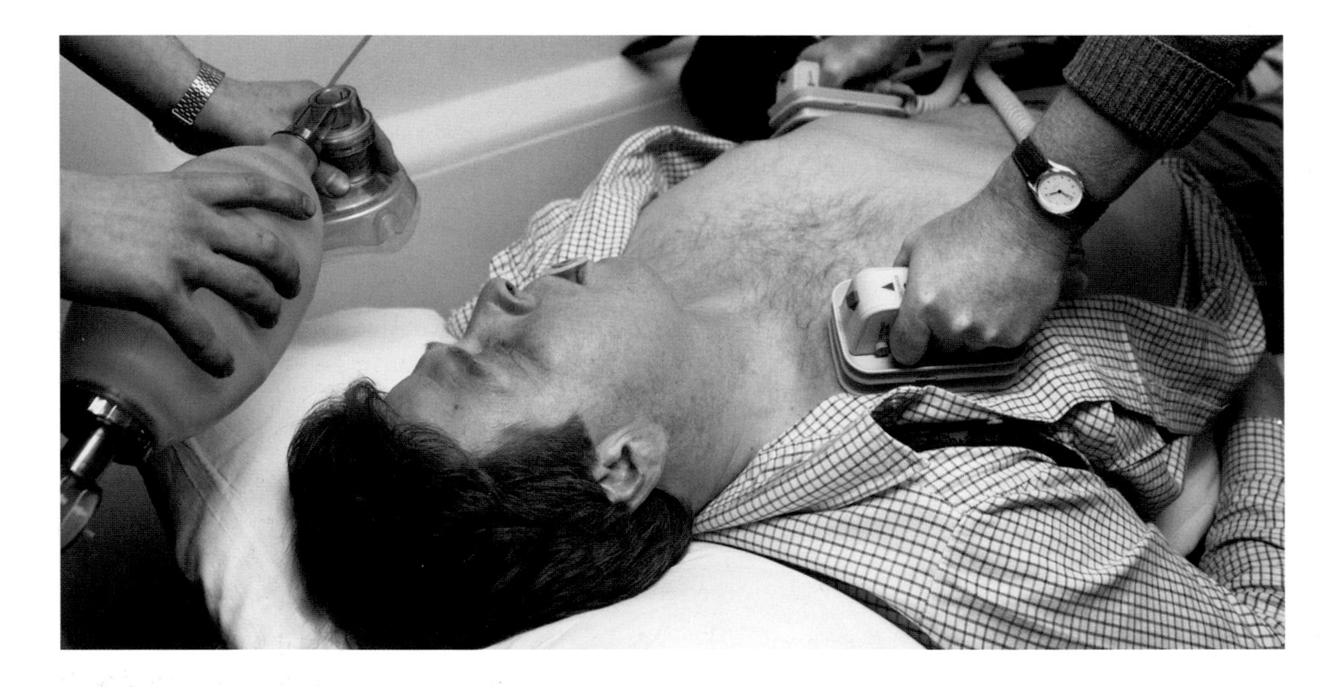

An indirect lightning strike It is also possible to be struck indirectly, via objects close by. For example, if a person touches a conducting object that is itself struck by lightning (a pipe system, the wall of a cave), a considerable amount of current can travel through the body and be extremely dangerous. It is also very dangerous to take shelter under a tree: since sap is not a good conductor, the presence of a person nearby can divert the current toward the path of least resistance, causing a lateral flash that passes through the person before reaching the ground. Such lateral flashes can also be transmitted from one individual to another. Consequently, people who have the bad luck to be out in the open when a storm hits need to avoid being too close to each other so as not to be part of a group lightning strike. Finally, a ground strike, caused when the flash's current spreads in all directions after hitting the ground, can also reach a person who is nearby, but these do not cause fatal injuries. However, herds of cattle and flocks of sheep can be mortally struck by a single bolt of ground lightning because, in their case, the current passing from their front hooves to their back hooves (or the opposite) also passes through the thorax and heart.

^ A cardiac defibrillator being used on a patient.

Household lightning

It is not necessary to reach the same levels of charge as lightning in order to electrocute someone. Contrary to popular belief, the main factor determining the severity of an electric shock is not, in most cases, the voltage of the electric charge but the intensity of its current (amperage). In practice, it is entirely possible to electrocute oneself with the weak voltages in household use (120 and 240 volts)! From the moment when the body becomes an integral part of an electric circuit with a current capable of stimulating the nervous system and damaging internal organs, there is a risk of fatal shock. For example, the current that powers a 7.5-watt, 120-volt lightbulb is more than enough to electrocute a person if it travels from one hand to the other through the chest (see Figure 3).

The threshold for the detection of a current is 1 milliampere (mA) and a few more milliamperes are enough to cause an instinctive recoiling movement. When it reaches a threshold of 16 mA, the current causes an involuntary contraction of the muscles that can prevent a person from letting go of the object through which the electricity is circulating. This prolonged contact is dangerous if the current exceeds 20 mA, as its passage through the chest for a prolonged period can cause death by paralysing respiratory functions. These levels of current are very dangerous when a person is in a humid place or has wet hands, for example, since the human body's electrical resistance in these circumstances can drop from 100,000 to 1,000 ohms. According to Ohm's law, voltage (in V) is equal to resistance (in Ω) multiplied by the current (in amperes). The current associated with a voltage of 120 volts has, in a dry environment, a barely perceptible intensity of 0.0012 A, or 1.2 mA (120V/100,000Ω), whereas in wet conditions,

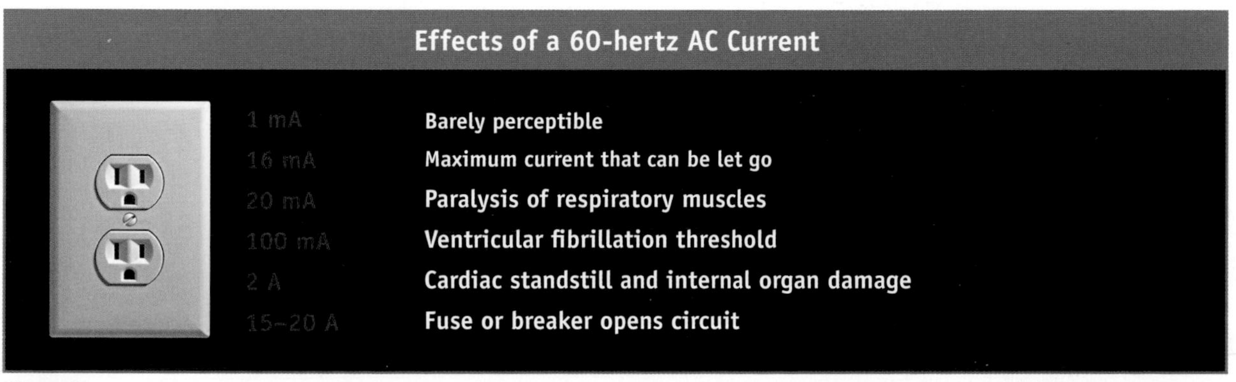

Figure 3

the same voltage generates a current intensity of 0.12 A, or 120 mA (120V/1,000Ω), enough to cause death. It is therefore not surprising that a large majority of electrocutions at weak voltages take place under wet conditions.

Stronger currents are of course still more dangerous, mainly because they can cause ventricular fibrillation, a very serious state in which the irregular contraction of the heart muscle walls keeps the blood from circulating and nourishing organs. This ventricular fibrillation is the main cause of death from electric shock. Only cardiopulmonary resuscitation to restart blood circulation in the brain and quick intervention using a cardiac defibrillator can prevent a fatal outcome. However, once the current exceeds 1 A, the damage to the cells is permanent, especially in the heart. As can be seen, the current needed to electrocute a human being is quite weak in relation to the currents normally found in a home, protected by 15 A and 20 A fuses, these being nearly 1,000 times more than the intensity needed to cause considerable damage.

The same electrical principle is at work in the electric gun commonly known as the Taser, invented by NASA engineer Jack Cover. Taser is an acronym derived from the science-fiction novel *Tom A. Swift and his Electric Rifle*. More than 450,000 of these guns have been manufactured to date and are used by police forces all over the world. The weapon fires two probes with sharp tips that hook onto the victim's clothes, transmitting a current of 2 mA with a potential of 50,000 volts. This paralysing electric discharge causes temporary loss of muscle coordination.

Unusual yet dramatic, the kinds of death discussed here remind us of the unpredictable nature of death and the vulnerability of our body in the face of the physical, chemical and electrical forces we come into contact with every day.

< Policeman armed with a Taser gun.

> A depiction of a god of thunder, Tenjin temple, Kamakura period.

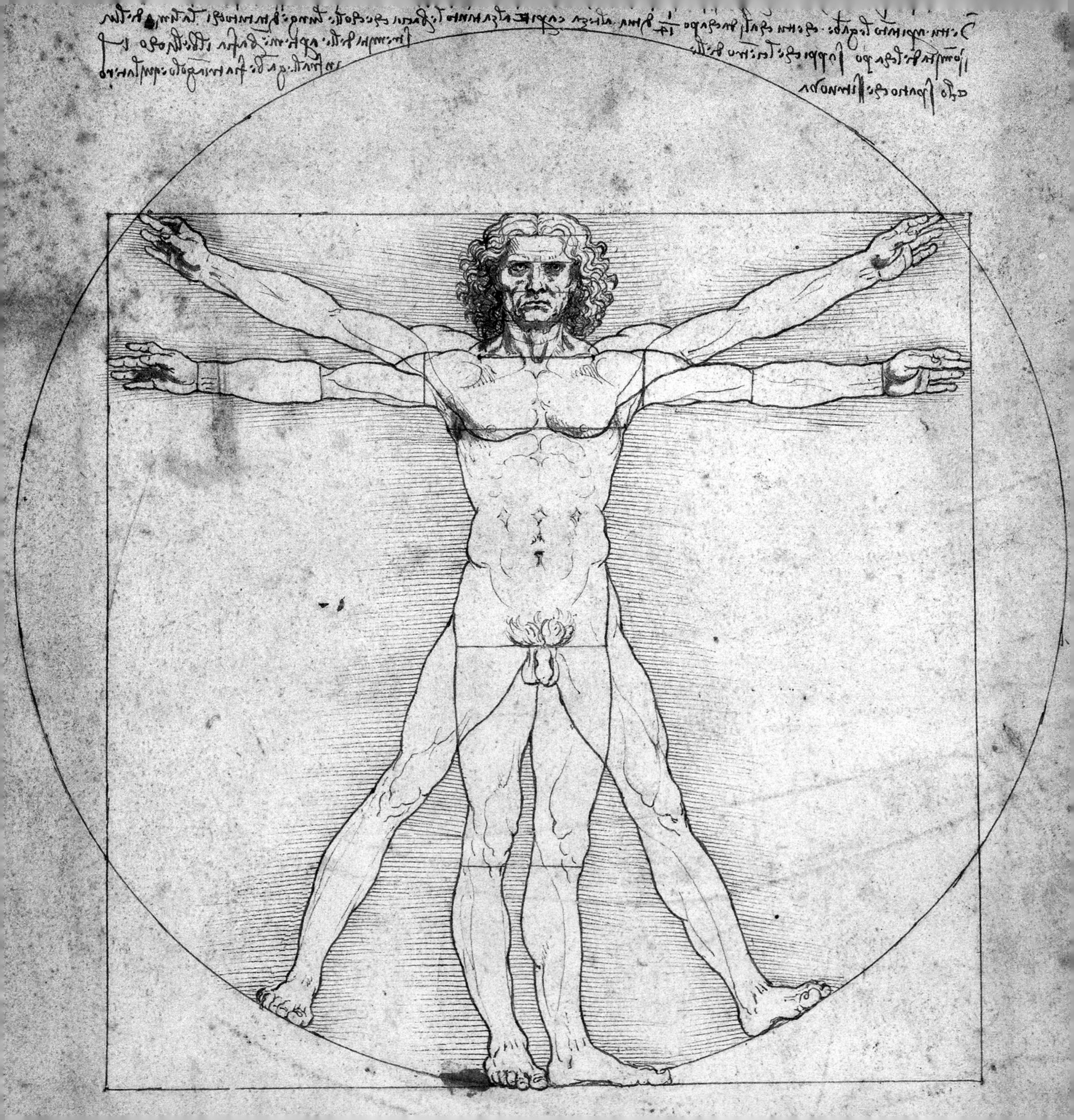

Chapter 10

Postmortem Events

From my rotting body, flowers will grow
and I will be a part of them and that is eternity.
Edvard Munch (1863–1944)

The events that occur after death fit into a much broader context involving the recycling of energy and matter out of which our universe is made and which we, as living beings, are part of. Our planet is a complex example of a large-scale ecosystem made up of living organisms. Formed around 5 billion years ago when a gaseous cloud and dust from the explosions of giant stars came together and condensed, Earth has retained essentially the same composition since then: each atom in the world around us today comes from the dust of original stars. Despite the colossal size of the universe and the "astronomical" number of galaxies, we live in a closed environment to which nothing is added, except for solar energy and some 16,000 tons of meteorite dust that falls to earth every year (this amount is negligible, however, given that the weight of the planet is around 13×10^{24} pounds/6×10^{24} kg). Every rock, every organism, whether bacterial, plant or animal in nature, is formed from these first atoms, which have been constantly recycled over billions of years. Of course, human beings are no exception to this rule, and the atoms that make up our body have undergone the same recycling process: among the average of 7×10^{27} atoms in our organism—hydrogen, oxygen, carbon and nitrogen alone make up 99% of all of these atoms (see Figure 1)—it is statistically possible that in the past some of these atoms briefly belonged to a tree, a bacterium and, who knows, even a dinosaur!

< *Vitruvian Man*, a study of the proportions of the human body by Leonardo de Vinci.

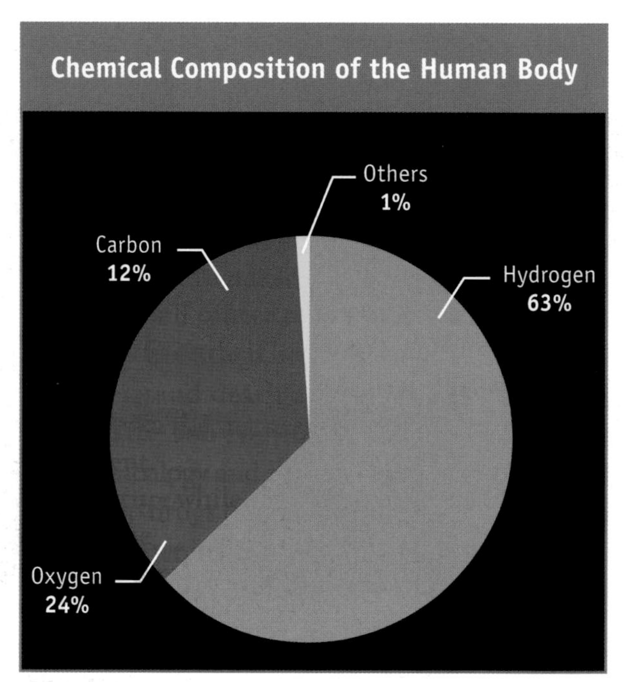

Figure 1

Of course, the same recycling logic applies to our current bodies. As the Bible says so well, "for dust thou art and unto dust thou shalt return" (Genesis 3:19), and whether we believe in the survival of the soul or not, one of the only certainties of life is that the collection of atoms we call the human body will be dismantled after our death. Far from being morbid, this idea of recycling can, on the contrary, be a source of some comfort, giving us the feeling that our death will bring about a redistribution of these elements so that other organisms, whether plants or animals, can grow and perpetuate the magnificent adventure of life on earth. We can understand those more sensitive readers who will prefer to avoid the description of how the human body decomposes; for the more curious, however, this is a truly fascinating subject that shows us death from a different angle, while also putting our presence on earth into a more global perspective.

Algor, livor, rigor

Despite its being natural and essential, the decomposition of a human being is not a very attractive sight. In fact, it is likely that underlying many funeral rites is our desire to hide from sight (and from the sense of smell) the changes the body undergoes after death. Indeed, whether the dead body is buried, cremated or exposed to birds (in Zoroastrianism), all these traditions share the perception of the body's deterioration as a revolting state, with the emphasis instead being placed on allowing the soul to go free.

The process of human decomposition begins very quickly, about 4 minutes after vital functions stop. However, the results of this process do not usually become apparent—

and in such a way that it likely would be traumatic to anyone not familiar with the putrefaction of organic matter — until a few days after death. Nevertheless, before arriving at this stage, a corpse undergoes a number of typical changes very familiar to murder mystery enthusiasts because of their important role in estimating time of death.

One of the first changes observable in a dead person is the appearance of postmortem lividity (livor mortis), which is a discoloration of the skin resulting from the pooling of blood in the lower portions of the body due to gravity. This accumulation occurs because blood stops coagulating shortly after death, owing to the massive release of anticoagulant enzymes (fibrinolysins) by the blood vessels. Lividity starts to develop as soon as blood circulation stops and peaks 12 hours after death. The slightest pressure on the blood vessels prevents the accumulation of blood at the point of contact, which means that the location of postmortem lividity is strongly influenced by the position of the body at the time of death, a very useful characteristic that helps forensic pathologists determine whether the body has been moved. For example, if the person died lying on their back, lividity will appear in the lower part of the torso and limbs as well as in the earlobes. If the blood is distributed in this way but the body is found face down, there is something fishy!

^ Believers watching the cremation of guru Sant Rama Nand in the Indian village of Bahwan, June 4, 2009.

A second significant change caused by death is the gradual cooling of the body (algor mortis). Its temperature gradually decreases to match that of the surrounding environment. The rate at which the body cools is strongly influenced by a wide array of environmental factors (clothing, amount of body fat, etc.), but the time of death can usually be estimated by calculating a drop of 4°F (2°C) during the first hour and 2°F (1°C) for each hour that follows.

The most intriguing change in the body that occurs in the hours immediately following death is postmortem rigidity (rigor mortis). This strange phenomenon, which usually begins 2 to 3 hours after death, is caused by the involuntary stiffening of the muscles. This rigidity first affects the muscles of the face and neck and then gradually progresses to the lower limbs. After 12 to 18 hours, the person is literally "stone dead," and this state may persist for up to 3 days after death. The mechanisms responsible for rigor mortis long remained a mystery and gave rise to several bizarre beliefs. For example, physicians in ancient Greece and Rome believed that it could raise the dead person to a sitting position. In reality, even though the muscles become rigid, this does not mean they can contract in any purposeful way (see box on p. 225).

^ Human remains

Dropping dead

The fascination with rigor mortis throughout history is understandable, as it is only recently—now that the biochemical mechanism controlling muscular contraction has been explained—that it has been possible to figure out what factors are responsible for this characteristic of death.

Whereas living cells use oxygen to make ATP, their main source of energy, the cells in an organism that has just died are forced to find an "emergency route," and the glucose reserves in the muscles are used to form ATP. This mechanism, however, has limited effectiveness, not only because ATP is completely depleted a few hours after death, but also because it leads to the production of metabolic wastes, which acidify muscle cells and change the function of proteins normally involved in muscle contraction. Under normal conditions, these proteins (actin and myosin) interact only when they receive the order via a signal from the nervous system. In the muscle cell of a dead person, the cell's acidity and the depletion of ATP remove this restriction, which leads to an abnormal interaction among the muscle fibers and the subsequent stiffening of the muscle. Even though this interaction between actin and myosin is very strong and continues until the process of decomposition gets underway, it is nonetheless not sufficient to produce real contraction and coordinated muscle movement, since that process absolutely requires ATP to be present in the cell. Rigor mortis is thus primarily caused by the total depletion of ATP several hours after death. This explains why people who die following intense muscular activity (and who have thus used up most of their glucose and ATP reserves) are usually more rapidly affected by post-mortem rigidity.

Rigidity must not be confused with cadaveric spasm, a rare form of muscular rigidity occurring at the very moment of death. The cause of this phenomenon remains unknown, but it is usually associated with violent deaths.

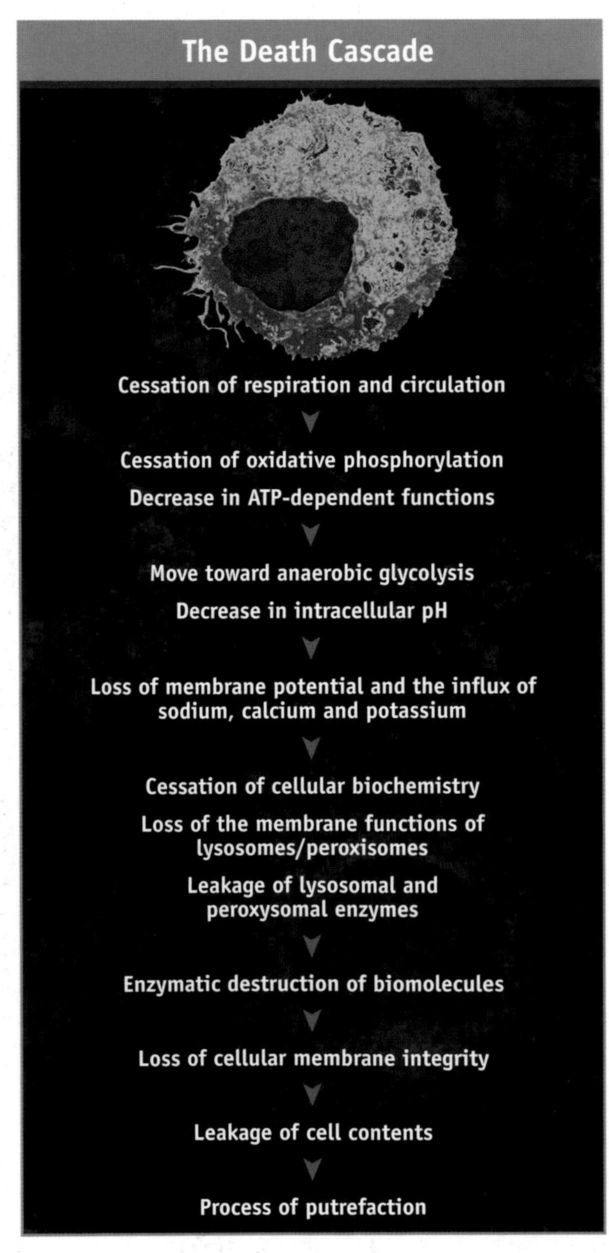

Figure 2

Post-mortem self-destruction

Another consequence of the lack of oxygen after death is the beginning of what is really a process of cellular self-destruction in the body's various organs (see Figure 2). Just like rigor mortis, this process, called autolysis (self-digestion), is also triggered mainly by the acidification of the cell, which is caused by lactic acid produced by the combustion of ATP in the absence of oxygen. This acidification triggers a cascade of events that irreversibly alter the very structure of the constituents of the cells. This is the swan song for what was, until then, a wonder of organization and order. First the membranes of the various cellular compartments dissolve and become permeable, then certain electrolytes, notably potassium, are expelled from the cells and may accumulate abnormally in some tissues (for example, potassium in the vitreous humor of the eye is used to estimate the time of death). Next a flood of degradation enzymes, which are normally stored in specialized compartments separate from the rest of the cell, are released and cannibalize the cell's main constituents (proteins, lipids, DNA), which have now become vulnerable to their action. The organs involved in digestion (the pancreas and intestines) are particularly affected as they contain large quantities of these enzymes. Some days after the death of the entire organism, the cells

> One of the plastinated human bodies in Gunter von Hagens's anatomical exhibition.

in turn surrender, leaving the body in a slightly acidic state, without oxygen and with a rich supply of nutritive substances that are no longer protected by the structure of the cells nor by any immune defense. An ideal ground for opportunistic microorganisms!

If there is no embalming or rapid cremation, the first noticeable sign of decomposition is the appearance, about 48 hours after death, of a greenish-colored blotch in the lower right region of the abdomen (iliac fossa), adjacent to the first part of the colon (cecum). This discolor-

The smell of death

It appears that the ability to detect the smell of death is a fundamental characteristic of all living beings. Recent experiments show that even ants and cockroaches recognize certain molecules produced when a member of their species dies and avoid remaining in the same place (a very useful instinct, as this odor could indicate the presence of a successful predator in the area).

In humans, the stench produced by decomposition is mainly due to hydrogen sulfide (H_2S), a gas with an odor reminiscent of rotten eggs or excrement (depending on the person). Two evocatively named products of protein degradation, cadaverine and putrescine, also contribute to the nauseating odor of a decomposing corpse. These molecules are in fact used to train cadaver-sniffing dogs. These dogs all have an impressive sense of smell, which means they can even detect corpses under water. For example, one such dog succeeded in detecting the odor of a swimmer drowned more than a year earlier in Lake Geneva in Switzerland even though the body was 148 feet (45 m) below the surface. The best breed of cadaver dog is the bloodhound.

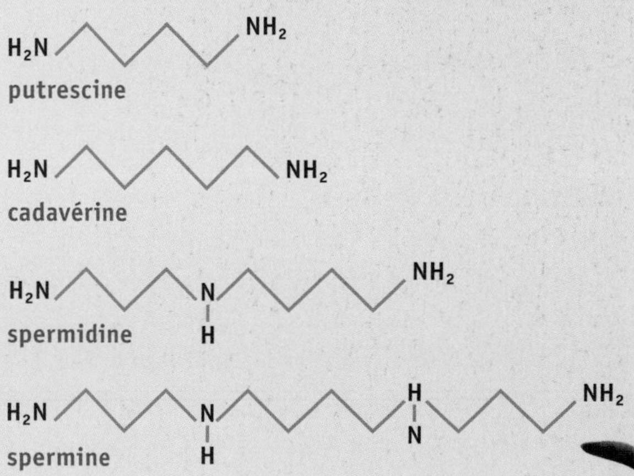

H_2N NH_2
putrescine

H_2N NH_2
cadavérine

H_2N N H NH_2
spermidine

H_2N N H H N NH_2
spermine

ation is caused by the rapid proliferation of the many bacteria in this part of the intestine (billions per ounce of tissue) that generate gases such as hydrogen sulfide (H_2S). This gas reacts with iron in the blood's hemoglobin to form sulfhemoglobin, a greenish-colored byproduct. Under the growing pressure of the resulting gas, these blotches then spread to other parts of the body (thorax, head, arms and legs) and become darker and darker. At the same time, the intestinal bacteria are joined by a sizeable contingent from outside and begin to spread throughout the whole organism via the blood vessels. This phenomenon shows on the skin as marbling, due to the coloration of surface veins. These changes in skin color are accompanied by the formation of blisters (phlyctenule) filled with the gas and liquid byproducts of decomposition. These blisters can be as large as 8 inches (20 cm) across and expose large areas of the dermis when they burst.

The gas produced by the proliferation of bacteria exerts more and more pressure on the body's outer wall with time, causing a dramatic ballooning effect that nearly doubles the corpse's volume. Of course, the abdomen is particularly inflated given the large quantity of bacteria there, but other regions are also affected, notably the head: the eyeballs are pushed out of their orbits, the lips swell and the tongue hangs out. These microbial gases are also responsible for the nauseating smell that accompanies the decomposition process, which is due to a number of its components, including hydrogen sulfide, certain fatty acid derivatives (butyric and propionic acids) and the products of protein degradation (cadaverine, putrescine, see box on p. 229).

While a large portion of these gases can escape through the natural orifices (mouth, nostrils, rectum and vagina), sometimes the skin cracks under the high pressure they exert or as a result of handling the corpse. There is a striking example of this in the description of the death of William the Conqueror (on September 9, 1087) as reported by historian Orderic Vitalis (*History of Normandy, Book VII*, c. 1140):

^ Human remains infested with maggots and worms.

> However, as we lowered the body into the coffin, and tried to bend it, as the coffin was too small thanks to the workers' clumsiness, the belly, which was very fat, burst open and an intolerable stench struck the people around it as well as the rest of those present. In vain, large amounts of smoke from incense and other aromatic substances rose out of the censers: it could not disguise the horrible stench it exuded. This is why the priests hurried to end the ceremony and, terrified, withdrew immediately to their dwellings.

And we can understand why... The overabundance of gas is a direct consequence of the intense metabolic activity of the microorganisms that have taken over the corpse. To feed themselves, the bacteria release a multitude of enzymes specialized in breaking down complex molecules (proteins, polysaccharides) to reduce the size of these molecules so they can absorb them. This process is no doubt essential for the growth of the bacteria, but it generates many smelly gases as metabolic wastes. However, aside from producing nauseating gases, this bacterial breakdown plays a key role in the decomposition of the body by completely liquefying and dissolving the tissues.

A large number of factors determine how quickly the decomposition process takes place and what it looks like. Thus, bacterial proliferation will be much faster in someone who dies from an infection or an open wound. Conversely, a person murdered with a violent poison will decompose more slowly, since the molecules will also be toxic for the microorganisms that cause decomposition. The ambient temperature, the level of humidity and the place where the corpse is located obviously have a determining influence on the process. As a general rule, 1 week of decomposition in the open air equals 2 weeks in an aquatic environment and 8 weeks in the ground.

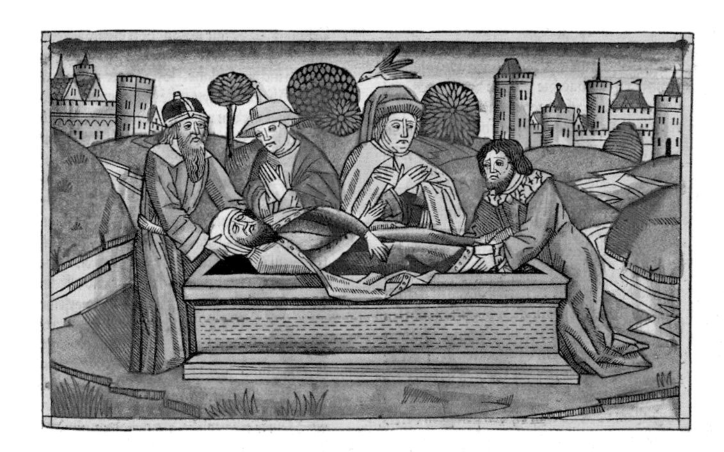

^ "Burial of Josias," from the *Biblia Sacra Germanaica*, artist unknown.

Insects join in

> "I heard a fly buzz when I died."
> Emily Dickinson

This famous line is prophetic, given the great attraction corpses have for many scavenger insects, particularly when the corpse is in the open air. Since the order in which scavenger insects colonize a body follows a clear sequence, it is possible to correlate the presence of some with the time and place of death, information that is often key in murder investigations. This investigation strategy is not new. In the 13th century, following the murder of a Chinese farmer who had been struck with a sharp object, all the suspects were summoned to the scene of the crime with their sickles. One of these sickles attracted a large number of flies, owing to the presence of blood residues, which forced its owner to confess to the murder.

However, it was only with the 1894 publication of the famous book *La Faune des cadavres: Application de l'entomologie à la médecine légale* (The fauna of cadavers: The application of entomology to forensic pathology) by Jean-Pierre Mégnin (1828–1905) that the way insects colonize a decomposing body was described in detail, giving birth to what today is called forensic entomology. Mégnin observed not a general rush of insects to the decomposing flesh, but instead what was really a well-ordered procession, during which eight distinct "squadrons" followed one another, depending on whether they preferred fresh flesh or flesh in various stages of decomposition. The first insects are usually diptera such as Muscidae (the house fly), Calliphoridae (bluebottle and greenbottle flies) and Sarcophagidae (flesh fly). These flies like to lay eggs on very fresh corpses and can appear just a few minutes after death, well before the process of decomposition has really begun. They lay their eggs in the natural orifices of the body or in open wounds, and their larvae, called "maggots," then develop by feeding on the decomposing body to form intermediary pupas and finally adult flies. After these first grave diggers come different varieties of Coleoptera, flies and beetles with highly evocative names (*Drosophila funebris, Necrobia violacea, Necrophorus humato*, among others). While some of these insects feast directly on the corpse, other even more opportunistic varieties use the scavenger insects themselves as their food source. The speed with which each of these species arrives and multiplies is strongly influenced by the climatic conditions surrounding the corpse as well as by its "availability." However, even though the speed of decomposition is considerably reduced when the body is embalmed and buried at some depth in an airtight coffin, the outcome is never in doubt: the human body, be it buried, cremated or embalmed, ultimately returns to a state of dust, giving back to the planetary ecosystem precious atoms that will be used to continue the adventure of life on earth.

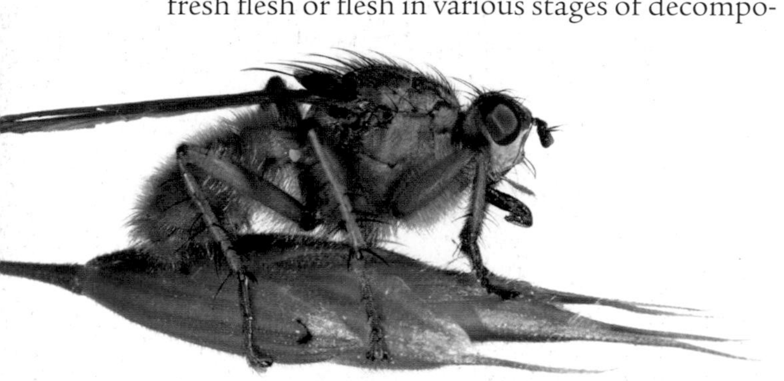

^ Yellow dung fly (*Scathophaga stercoraria*)

> The Milky Way

Natural mummies

Under certain specific climatic conditions, the way the decomposition process unfolds is changed; the result is incomplete decomposition of the corpse. For example, Ötzi, a human being buried under a layer of ice for some 5,000 years, was found by accident in 1991 in the mountains between Italy and Austria and was relatively well preserved by freezing.

The much "younger" Tollund man was found in 1950 in a peat bog in Denmark. This natural mummy is particularly striking because of the astonishing clarity of his facial features. Yet analyses show that he died around 400 years BCE! This degree of conservation is due to the acidity of the water in peat bogs; combined with a cold environment lacking in oxygen, this acidity causes the skin to dry out and tans it naturally.

One of the most dramatic anomalies of decomposition is the formation of adipocere, a substance with a waxy consistency found on the surface of some corpses. This term, derived from the Latin *adeps* ("fat") and *cera* ("wax"), was first used by French chemist Antoine-François Fourcroy to describe the substance mid-way between fat and wax that he had seen on the corpses of exhumed children when the Saints Innocents Cemetery was closed down in Paris in the 18th century. Intrigued by the composition of this substance, Fourcroy and his colleagues later showed that it was chemically similar to soap!

We now know that the appearance of adipocere is due to the breakdown of adipose tissue by certain anaerobic bacteria (*Clostridium perfringens* in particular) , which releases free fatty acids so they can interact with certain ions released during the initial autolysis of the body. Under optimal conditions—a humid, alkaline and oxygen-poor environment—free fatty acids can combine with sodium or potassium to produce solid compounds, somewhat like the reaction that produces soap. When a corpse has undergone slow putrefaction, this "soap" is mostly formed from the sodium found abundantly in body fluids and has a softer consistency, a little like cheese. When putrefaction occurs more quickly, the release of potassium following cell autolysis leads to the formation of a harder substance resembling candle wax. Naturally, bodies containing a larger proportion of fat, such as children, women and the obese, are more likely to be covered in adipocere.

The transformation of body fat into adipocere is, however, a rare phenomenon, mostly found on bodies buried in humid ground or on victims of drownings. However, when it is present, adipocere causes the putrefaction process to slow down dramatically thanks to the antibacterial action of this "soap." For example, the analysis

< Tollund man

of the mortal remains of a child dating from the Roman era revealed a body coated in adipocere to be extremely well preserved more than 1,600 years after death.

Sokushinbutsu: Self-mummification

A custom originating in northern Japan, *sokushinbutsu* refers to a process through which certain Japanese monks voluntarily mummified themselves. This kind of death stems from an extreme interpretation of Buddhism, a philosophy according to which the perceived world is an illusion masking the existence of a supreme non-material being separate from the physical world. The first stage of mummification for these monks involved eating only grains and nuts for 1,000 days while engaging in as much physical exercise as possible. At the end of this "diet," body fat was reduced to a minimum, thus eliminating one substance that would normally decompose. For the next 1000 days, this diet was replaced by a food supply strictly limited to pieces of bark and roots with the addition, toward the end of this period, of a tea made from the sap of the urushi tree (*Toxicodendron vernicifluum*), a very toxic substance normally used as lacquer, and spring water from Mount Yudono. We now know that this water contains abnormally high concentrations of arsenic that, when combined with the lacquer,

^ Mummified body of a Buddhist monk

Embalming death

The ancient Egyptians had a decisive influence on the development of techniques designed to preserve the integrity of the body after death. In their society, death was deemed to be neither an ending nor a beginning, but rather an extension of life on earth. In order to reach the afterlife and enter the kingdom of Osiris, god of the dead, the body of the deceased had to be preserved to ensure that his soul would live on.

Already in practice since the Third Dynasty (2800 BCE), mummification peaked in the Eighteenth and Nineteenth dynasties (1550–1070 BCE), a period marked by the reign of Ramses II.

Creating a mummy was a highly complex process, and we have the fortune of knowing about it thanks to the observations of the Greek historian Herodotus. Lasting 70 days, or the length of the eclipse of Sirius, the procedure consisted of first liquefying the brain using a bronze rod inserted into the nasal cavity (ethmoid bone) and then filling the cranium with various resins having antiseptic properties.

^ Anubis preparing Sennedjem's mummified body.

> Then, using a sharp Ethiopian stone, they cut open the body, took all of the intestines out of the abdomen, washed it out with palm wine, sprinkled it with ground perfumes and finally sewed it up again, after having filled it with pure crushed myrrh, cinnamon and other perfumes, with incense being the only one not used.
>
> Herodotus, *Histories, Book II*, p. 86–87.

The body was then dehydrated for 50 days with natron, a crystalline deposit left behind when the water level dropped in the spring in certain lakes between Cairo and Alexandria and containing (among other substances) carbonate and bicarbonate of soda, which absorbed the moisture in the tissues. Lastly, the body was filled with substances intended to absorb any remaining liquid and wrapped in linen strips soaked in the gum or resin of conifers. From the New Kingdom on, the face and shoulders were covered by a mask, the most famous being that of Tutankhamen, made of solid gold.

Modern embalming

Embalming as practiced today does not aim to achieve the lengthy preservation of the body desired by the Egyptians. The purpose is intead to preserve the body for the time it takes to arrange burial rites and also to prevent the spread of diseases and improve the appearance of the deceased.

The embalming procedure, sometimes called thanatopraxia (from the Greek Thanatos, god of death), can vary considerably depending on the state of the corpse and the causes of death, but generally it involves draining the blood from the body by injecting several pints of a sterilizing liquid containing (among other substances) formaldehyde (from 5% to 35%) and ethyl alcohol (from 9% to 56%) into the carotid or femoral artery. An incision is made in the corresponding veins (jugular or femoral) to allow the body fluids to flow out. The formaldehyde in this liquid is a molecule with strong disinfecting power, and it also interacts with cell proteins to cause chemical fixation of the tissues; this fixation is what makes the skin of an embalmed person feel so firm. A coloring agent may also be added to this solution to create a pink shade, closer to how the person looked in life. The gases and liquids in the abdominal cavity are also removed and replaced by preservative agents. Although this method of embalming enables the deterioration of the body to be postponed considerably, it is only temporary, since microorganisms in the air and ground will eventually cause total decomposition after burial.

< Funeral mask of Tutankhamen

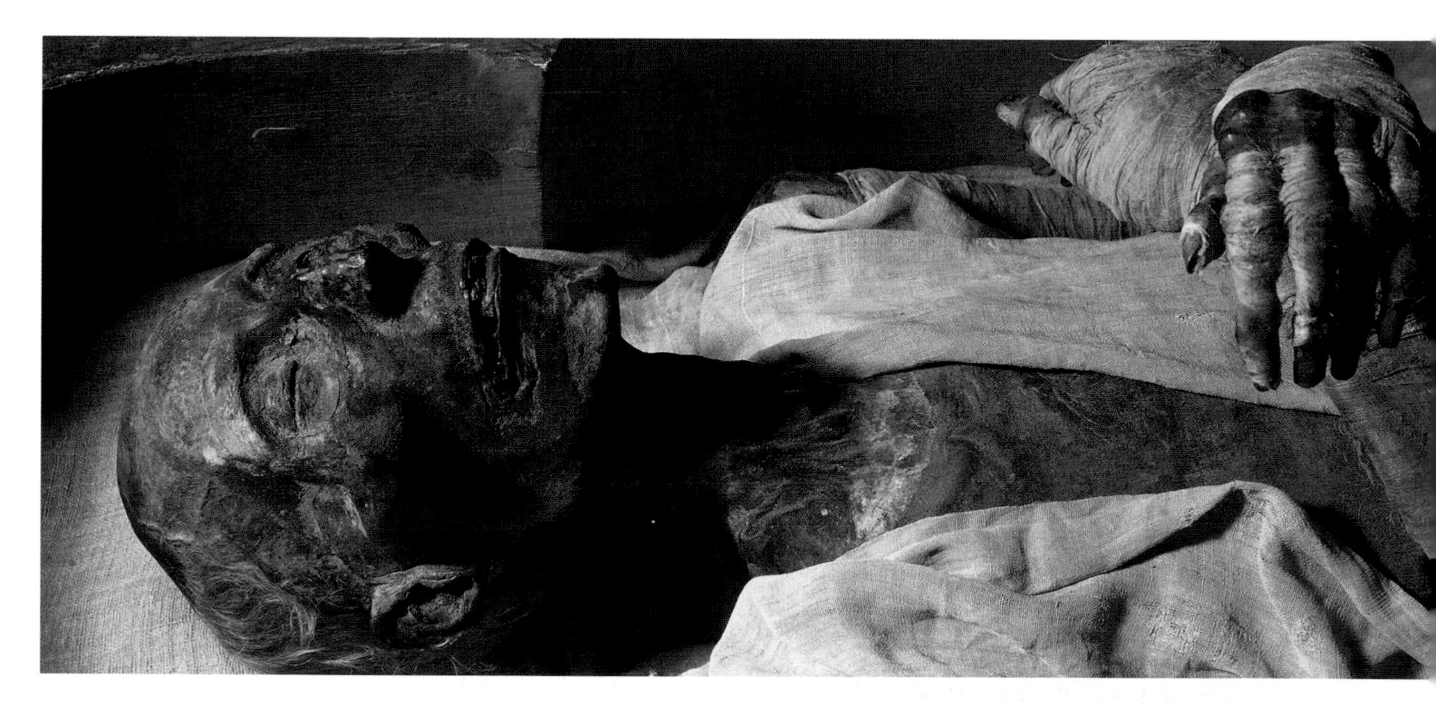

created a toxic and sterile environment inside the body and, hence, reduced the risk of decomposition after death by both bacteria and insects. Finally, at the end of this process, the monks retired into such a small underground chamber that they could only sit in the lotus position to meditate, linked to the external world only by a breathing tube and a bell they rang once a day to show they were still alive. When this bell fell silent, the tomb was sealed for a 1000 days, after which the bodies were exhumed. Those monks who had not decomposed were raised to the rank of Buddha. These mummies can still be viewed today in some temples in Japan.

Of all of our fears and uncertainties about death, the processes described in this chapter are, for many people, among the most disturbing. Instead of being a source of anxiety, these phenomena should lead us to reflect on the finiteness of our existence and on the modest and ephemeral nature of our presence on earth. Are modesty and humility not fundamental human values, common to many cultures?

It took the genius of one of the greatest poets in history, Charles Baudelaire, to take such an uninspiring theme and turn it into a love song!

^ Mummified body of Ramses II

Carrion

Darling, do you recall that thing we found
("A lovely summer day!" you said)
That noisome carcass where the path swung round
A sprawling pebble-covered bed.

Its legs raised like a whore's in lubric play,
It burned, oozing rank fetors there,
Shameless and nonchalant, it offered day Its belly.
Poisons filled the air.

The sun beat down on this putrescent mold
As if to fry it to a turn,
To give great Nature back one hundredfold
All she had gathered in her urn.

The skies watched that proud carcass, lax or taut,
Bloom like a flowery mass.
So pungent was the stench, my love, you thought
To swoon away upon the grass.

Horseflies buzzed loud over this putrid belly,
Whence sallied column and battalion
Of sable maggots, flowing like a mucose jelly,
Over this live tatterdemalion.

Waves seemed to rise and fall over this mass,
Spurting with crepitation,
As though this corpse, filled with breaths of gas,
Lived by multiplication.

This world uttered a curious melody,
Like waters, wind, or grains of wheat
That winnowers keep stirring rhythmically
In the broad baskets at their feet.

The forms, fading into a dream, grew fainter;
Here was a sketch of misty tone
On a forgotten canvas which the painter
Completes from memory alone.

Hiding behind the rocks, an anxious bitch
Stood, watching us with angry eye,
Poised to regain the olid morsel which,
Hearing us come, she had laid by.

— Yet shall you be like this ordurous blight,
You, too, shall rot in just such fashion,
Star of my eyes, sun of my soul's delight,

Aye, you, my angel and my passion.
Such you, O queen of graces, in the hours,
When the last sacrament is said,
That bear you under rich sods and lush flower
To molder with the moldering dead.

Then, O my beauty! Tell such worms as will
Kiss you in ultimate coition
That I have kept the form and essence of
My love in its decomposition.

Charles Baudelaire,
"Une charogne," in *Flowers of Evil, trans. Jacques LeClercq.*
Mt Vernon, NY: Peter Pauper Press, 1958 .s

> Pieter Claesz, *A Vanitas Still Life*

Chapter 11

Dying of Laughter

Beyond a shadow of a doubt, death, along with love, is the theme that has most inspired philosophical and poetic minds throughout human history. For in the end, isn't laughing about death the best way to triumph over it?

"All good things have an end. Except for sausages, which have two."

Jean-Marc Minotte, known as Jean l'Anselme

"What's the point in learning how to die? We do it very well the first time."

Nicolas de Chamfort

"No sooner does some great man's heart stop beating that they name a main street after him."

Eugène Labiche

"Do not take life too seriously. You will never get out of it alive."

Elbert Hubbard

"Nothingness is the universe without me."

Yves Scandel, known as André Suarès

"Health is a precarious state that does not bode well."

Jules Romains

"Time wrinkles human skin and smoothes tire skins."

Paul Morand

"Dying is both leaving the earth and entering it."

André Birabeau

"Living all the time gets deadly after a while."

Jacques Audiberti

"He's dead, but why should I go to his funeral since I'm sure he won't come to mine?"

Jacques Prévert

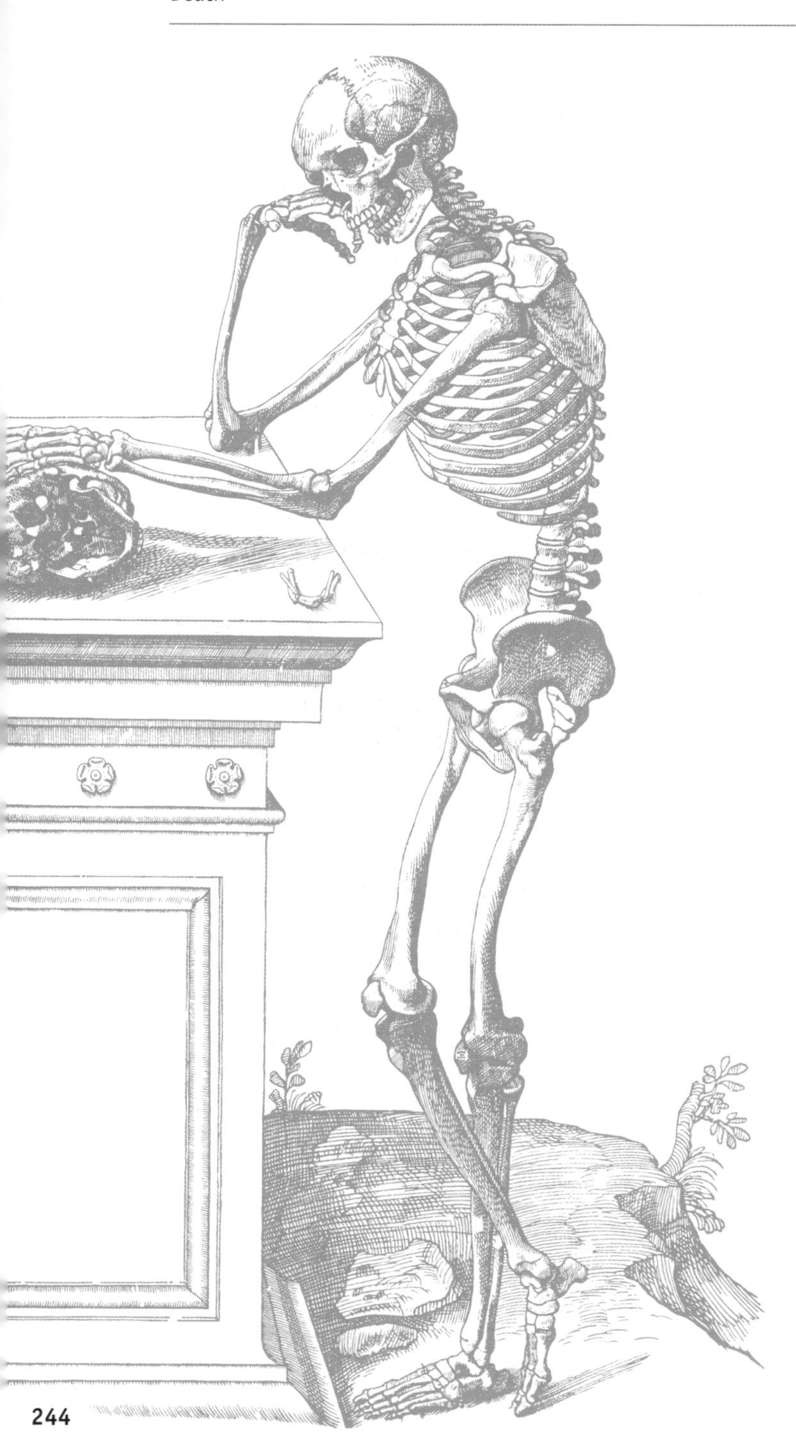

“Unable to stand the idea of death any longer, he killed himself.”

Claude Roy

“It’s the shadow of death that gives life its depth.”

Ingmar Bergman

“You have to live to be old, really very old, and even far too old. This gives you the pleasure, over the years, of seeing those who make fun of you buried.”

Jean Dutourd

“The best way to die is to be killed at 80 by a gunshot fired by a jealous husband.”

Francis Blanche

“Life: time on earth; Death: time in the earth.”

Jacques Kalaydjian, known as Jicka

“Hearty eaters dig their graves with their own teeth.”

Henri Estienne

“All human evil comes from a single cause, man’s inability to sit still in a room.”

Blaise Pascal

“Death won’t take me alive.”

Jean Cocteau

"We could give plenty of examples of useless expenses. Cemetery walls: those who are inside can't get out, and those who are outside don't want to get in."

Mark Twain

"The fact that he is dead in no way proves he ever lived."

Stanislaw Jerzy Lec

"The dead lamb is no longer afraid of the wolf."

Russian proverb

"The death rate would be very high if we stopped living when we have nothing more to say."

Abbé de Voisenon

"Sleep is borrowed from death to sustain life."

Arthur Schopenhauer

"He who is to die dies in darkness even if he is a candle-seller."

Colombian proverb

"Toughness and rigidity are the companions of death. Fragility and flexibility are the companions of life."

Chinese proverb

"The pope's corpse takes up no more space than that of the sexton."

Spanish proverb

"Old age is wonderful; it's a shame it ends so badly!"

François Mauriac

"Everyone must strive to die well. In the end we all turn to ash."

French proverb

"Death does not lie on the other side of the mountains; it is right behind us."

Russian proverb

"There are two things that cannot be stared at: the sun and death."

Russian proverb

"In the end, death is just caused by a lack of education, since it is the result of not knowing how to live."

Pierre Dac

"Some people die too soon. Many die too late. Very few die at the right time."

Friedrich Nietzsche

"Aside from disease, war, and death, how are things?"

Joseph Delteil

"I would like to die standing in a field, in the sun
Not in a bed with rumpled sheets
In the closed-off shade of shutters without so much as a bee."

Jean Ferrat

"Just one more minute and death would have come, But then a naked hand appeared and took mine.
Who gave back their lost colour to the days, the weeks?
Its reality to the vastness of human things. Just a movement,
This gesture while sleeping, gentle, brushing against me
A forehead resting against me in the night,
Big wide-open eyes
And everything seemed to me to be a field of wheat in this universe."

Louis Aragon

"Bah! All I'm leaving behind me are the dying."

Ninon de Lenclos (at the time of her death)

"As for dying, that's the last thing you want to do."

André Wurmser

"Man, this person condemned to death."

Jules Renard

"Death is so inevitable that it's almost a formality."

Marcel Pagnol

"Death is an idea of the living. What makes it appalling is the amount of living that the living put into it."

Paul Valéry

"You die, we do the rest."

Ad slogan for the American funeral industry

"We don't prepare ourselves for death. We detach ourselves from life."

Paul Claudel

"Diseases: death's way of checking the fit."

Jules Renard

"The stupidest phrase ever heard by any human ear is the one about 'singing tomorrows.' What other tomorrow is there but decay, dissolution and nothingness?"

François Mauriac

"Every man who is going to die consists of two men: the man he was, who keeps the man he is upright."

Henry de Montherlant

"Pain is a century and death but an instant."

Jean-Baptiste Grasset

"Death isn't the greatest mystery, life is."

Henry de Montherlant

"Philosophizing is learning how to die."

Michel de Montaigne

"Death? I should live so long!"

Jean Paulhan

"It is noble to learn how to die."

Epicurus

"Death does not impress me; indeed I myself am quite determined to die one day."

Jules Renard

"How many men die in a man before his death!"

Edmond and Jules de Goncourt

"Believe that each day is your last."

Horace

"Man accepts death, but not his time of death. He is ready to die anytime, except when it's his time to die."

Emil Michel Cioran

"If you want to live, you also want to die; or else you don't understand what life is all about."

Paul Valéry

"Life kills."

Louis Scutenaire

"Today me
Tomorrow you"

Inscription on the gate of a cemetery in Algiers

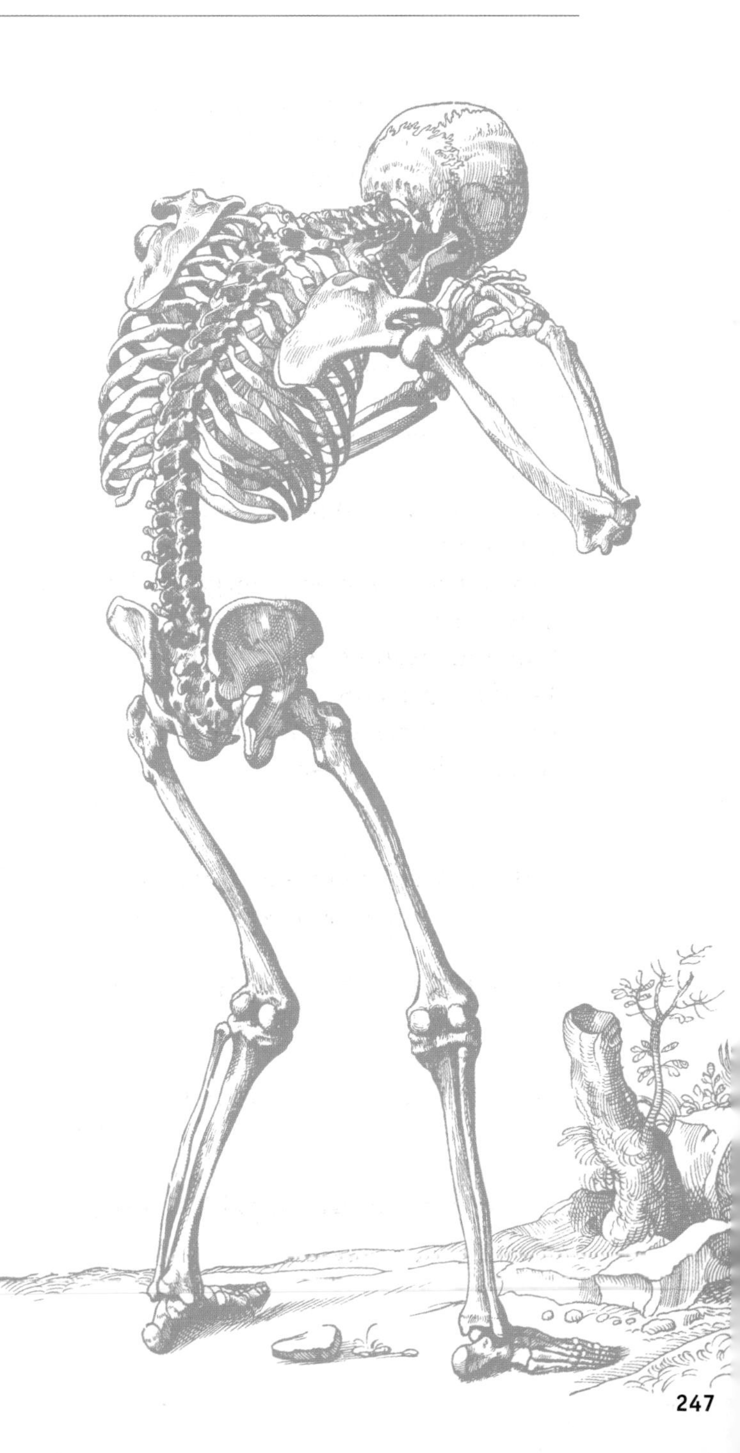

"To demand that an individual be immortal is to want to perpetuate an error endlessly."

Arthur Schopenhauer

"He shook hands with all the dead and joined the queue behind them."

Elias Canetti

"I told you I was sick!"

Inscription on a gravestone

"No matter which way you look, death is on the watch."

Sign in the Paris catacombs

"They were what we are
Dust, the wind's plaything
Fragile like men
Weak like nothingness."

Alphonse de Lamartine

"I see death at work in the mirror."

Jean Cocteau

"He let go of his last fear and died."

Elias Canetti

"The immortality of the soul was invented either because we fear death or because we miss the dead."

Gustave Flaubert

"Nothing but the fear of death keeps them alive."

Jules Renard

"Once you really look at it, death is not unpleasant to understand."

Jules Renard

"The death of one man is a tragedy. The death of millions of people is a statistic."

Joseph Stalin

"Death is kind; it delivers us from the thought of death."

Jules Renard

"Death is big... it's full of life."

Félix Leclerc

"Who dies not before he dies is lost when he dies."

Jacob Boehme

"To my husband, dead after a year of marriage." From his grateful wife.

Inscription on a gravestone in Père Lachaise cemetery

"You have your whole life to have fun in and eternal death to rest in."

François Rabelais, *Treatise on the Proper Use of Wine*

"He died with a great future behind him."

James Joyce

"No matter how high you climb, you always end up as ashes."

Henri Rochefort

"Life is pleasant. Death is peaceful. It's the transition that's troublesome."

Isaac Asimov

"Bed is the most dangerous place in the world: 80% of people die there."

Mark Twain

"When her husband died, she finally stopped feeling lonely."

Gilbert Cesbron

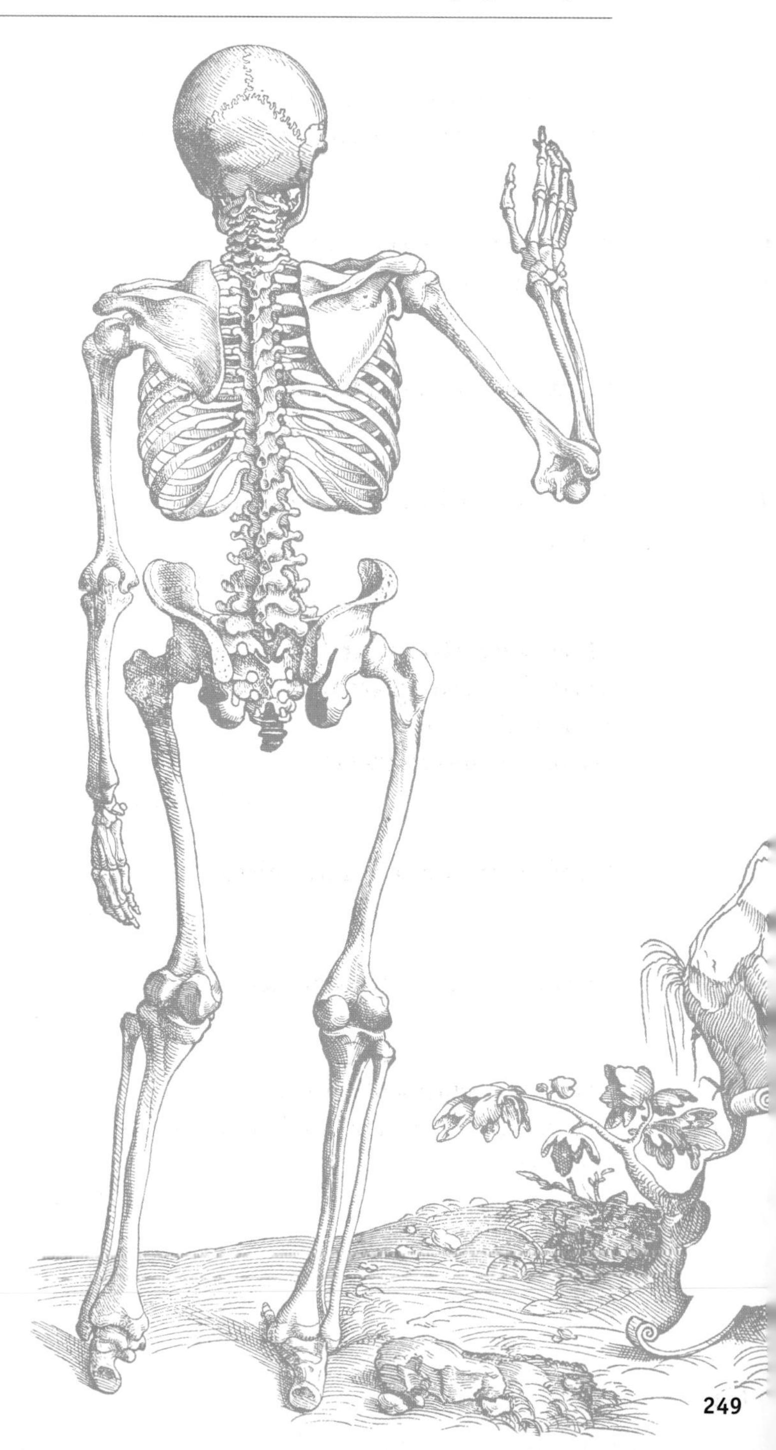

> Temple of Ryoan-ji (following page)

"Fear is suffering.
Fear is not accepting what is.
Fear only exists in relation to something.
It is the mind that creates fear.
Only self-knowledge can free you from death.
Self-knowledge is the beginning
of wisdom and the end of death."
Jiddu Krishnamurti

Thirty years ago, Richard Béliveau's friend and mentor, Ben Sulsky, had a serious heart attack, requiring several major surgiries. He came very close to dying. He decided then and there to make draconian changes in his lifestyle. Quitting smoking, daily physical exercise, a healthy diet, regular intellectual activity and philanthropic involvement—these were the tools he used to take charge of his life.

He learned to play tennis at 73 and golf at 80. He will be 87 this year. He is madly in love with life, an epicurean with a great sense of humor, a curious and iconoclastic mind and an exceptionally generous human being. Could it be that this love of life and his unselfish nature are the best ways to triumph over our fear of death?

Conclusion

Men live as if they were never going to die
And die as if they had never lived.
The Dalai Lama (1935–)

I do not fear death. I had been dead for billions and billions of years
before I was born, and had not suffered the slightest inconvenience from it.
Mark Twain (1835–1910)

While acute awareness of the ephemeral nature of existence is a fundamental characteristic of the human species, death remains nonetheless an individual ordeal that each of us copes with in our own way and to the best of our ability. There are no general "instructions" for dealing with death; our attitude toward death is instead a complex amalgam of emotions derived from the combined effect of our experiences, knowledge and genes, as well as their overall influence on our values and our perception of what life is. For some people, all of these factors contribute to make death the ultimate fear, an absolute evil impossible to imagine; for others, it is instead a normal event, viewed with serenity. While the fear of death is instinctive for all living beings and therefore perfectly normal, the terror it arouses, is simply a construction of the mind, an exclusively human phenomenon.

The major part played by human experience in the anxiety associated with death nonetheless offers the possibility of attenuating this fear, of making the most of our unusual intellectual faculties to come to grips with the inevitability of our aging and, ultimately, our death. Looked at this way, one of the major difficulties in accepting death very often stems from our lack of understanding of the phenomena at work in the appearance and maintenance of life. When we become aware of the extraordinary number of processes that are essential for survival and the immense obstacles that have had to be overcome to arrive

< Caspar David Friedrich, *Wanderer Overlooking the Sea of Fog* (c. 1818)

at the profusion of life as we know it today, we cannot help but be amazed to have had the opportunity to be born. Death is not an abnormal or absurd event. On the contrary, it is having the good luck to live that is something of a miracle.

Death is an essential prerequisite for life to continue and evolve. Just as the innumerable living beings that have preceded us have enabled us to be born, so it is our own death that will allow new generations to enjoy life in their turn. An immortal world would be doomed to remain static and fossilized, with the intrinsic limits on each individual preventing any real change.

We did not exist for almost the entire life of the universe, and in a few decades we will have disappeared once again, just like all the beings now alive on earth. But for a short moment—all too brief!—exceptional conditions came together to create a unique life, our own, that had never existed before and will never exist again. Instead of worrying about death, let's instead take advantage of this brief time on earth to get the most out of life and make the most of our good luck in having been able to be a part of this incredible adventure. Life is a sublime experience, even though it has no choice but to end in death.

Bibliography

The nature of the subject matter in this book required the consultation of a wide array of reference works, only a few of which are mentioned here.

Chapter 1

The Brain from Top to Bottom!, http://thebrain.mcgill.ca/

Linden, D.J. *The Accidental Mind: How Brain Evolution Has Given Us Love, Memory, Dreams, and God.* Cambridge: Harvard University Press, 2007.

"Conscience : les nouvelles découvertes." *La Recherche* 439 (March 2010).

Blanke, O., and S. Arzy. "The Out-of-Body Experience: Disturbed Self-Processing at the Temporo-Parietal Junction." *Neuroscientist* 11 (2005): 16–24.

Chapter 2

Ciccarelli, F.D. et al. "Toward Automatic Reconstruction of a Highly Resolved Tree of Life." *Science* 311 (2006): 1,283–1,287.

Powner, M.W. et al. "Synthesis of Activated Pyrimidine Ribonucleotides in Prebiotically Plausible Conditions." *Nature* 459 (2009): 239–242.

Lane, N. *Power, Sex, Suicide: Mitochondria and the Meaning of Life.* Oxford: Oxford University Press, 2006.

Kirschner, M.W., and J.C. Gerhart. *The Plausibility of Life: Resolving Darwin's Dilemma.* New Haven: Yale University Press, 2005.

Dawkins, R. *The Greatest Show on Earth: The Evidence for Evolution.* Free Press, 2009.

Chapter 3

Morin, E. *L'Homme et la Mort.* Paris: Éditions du Seuil, 1976.

Wright, R. *The Evolution of God.* Little, Brown and Company, 2009.

Hall, J. "Biochemical Explanations for Folk Tales: Vampires and Werewolves." *Trends in Biochemical Sciences* 11 (1986): 31.

Chapter 4

Fries, J.F. "Aging, Natural Death, and the Compression of Morbidity." New *England Journal of Medicine* 303 (1980): 130–135.

Colman, R.J. et al. "Caloric Restriction Delays Disease Onset and Mortality in Rhesus Monkeys." *Science* 325 (2009): 201–204.

Leslie, M. "Aging: Searching for the Secrets of the Super Old." *Science* 321 (2008): 1,764–1,765.

The Science of Staying Young, Scientific American, Special Edition (June 2004).

Chapter 5

Nuland, S.B. *How We Die: Reflections of Life's Final Chapter.* Vintage, 1995.

Zipes, D.P., and H. J. Wellens. "Sudden Cardiac Death." *Circulation* 98 (1998): 2,334–2,351.

Physicians' Desktop Reference, http://www.pdrhealth.com:80/home/home.aspx

World Health Organization, www.who.int/en

Emanuel, E.J., "Euthanasia: Historical, Ethical, and Empiric Perspectives." *Archives of Internal Medicine* 154 (1994): 1,890–1,901.

Chapter 6

Barry, S., and N. Gualde. "La Peste noire dans l'Occident chrétien et musulman, 1347–1353." *Bulletin canadien d'histoire de la médecine* 25 (2008): 461–498.

Kelly, J. *The Great Mortality: An Intimate History of the Black Death, the Most Devastating Plague of All Time.* Toronto: Harper Collins, 2005.

Engleberg, N., V. DiRita, and T. Dermody (eds.). *Schaechter's Mechanisms of Microbial Disease.* 4th edition. Lippincott Williams & Wilkins, 2006.

Shinya, K. et al. "Avian Flu: Influenza Virus Receptors in the Human Airway." *Nature* 440 (2006): 435–436.

Taubenberger, J.K., and D.M. Morens. "The Pathology of Influenza Virus Infections." *Annual Review of Pathology* 3 (2008): 499–522.

Neumann, G. et al. "Emergence and Pandemic Potential of Swine-origin H1N1 Influenza Virus." *Nature* 459 (2009): 931–939.

Chapter 7

Mead, R.J. "The Biological Arms Race: Evolution of Tolerance to Specific Toxins." *Proceediwngs of the Nutrition Society of Australia* 11 (1986): 55–62.

Appendino, G. et al. "Polyacetylenes from Sardinian Oenanthe Fistulosa: A Molecular Clue to Risus Sardonicus." *Journal of Natural Products* 72 (2009): 962–965.

Goldfrank, L. et al. *Goldfrank's Toxicologic Emergencies. 7th edition.* McGraw-Hill Professional, 7th edition, 2002.

"A Brief History of Poisoning", http://www.bbc.co.uk/dna/h2g2/A4350755

Chapter 8

Patrick, U.W. "Handgun Wounding Factors and Effectiveness." Quantico: Firearms Training Unit, FBI Academy, 14 July 1989. http://www.firearmstactical.com/pdf/fbi-hwfe.pdf

Chapter 9

Sanchez, L.D., and R. Wolfe. "Hanging and Strangulation Injuries." *Harwood Nuss' Clinical Practice of Emergency Medicine*, 4th edition. Lippincott Williams & Wilkins, 2005.

Pattinson, K. "Opioids and the Control of Respiration." *British Journal of Anaesthesia* 100 (2008): 747–758.

Chapter 10

Vass, A. "Beyond the Grave: Understanding Human Decomposition." *Microbiology Today* 28 (2001): 190–192.

Goff, M.L. "Early Post-mortem Changes and Stages of Decomposition." In Amendt J., Goff M.L., Campobasso, C.P., and M. Grassberger, eds., *Current Concepts in Forensic Entomology*, 2010, 1–24.

Amendt J. et al. "Forensic entomology." *Naturwissenschaften* 91 (2004): 51–65.

Department of Forensic Medicine, University of Dundee. "Postmortem Changes and Time of Death," http://www.dundee.ac.uk/forensicmedicine/notes/timedeath.pdf

Fiedler, S., and M. Graw. "Decomposition of Buried Corpses, with Special Reference to the Formation of Adipocere." *Naturwissenschaften* 90 (2003): 291–300.

Illustration Credits

Amélie Roberge: 16, 20, 21, 29, 32, 35, 36, 53ab, 55, 57, 59, 60, 63, 66, 101, 105, 112, 113, 116, 126, 148b, 150, 191, 195

Bridgeman Art Library: Arthur M. Sackler Gallery, Smithsonian Institution, U.S./The Anne van Biema Collection 198; Biblioteca Ambrosiana, Milan, Italy 181a; Bibliothèque Nationale, Paris, France/Archives Charmet 134a, 140; British Museum, London 77; Casa di Dante, Florence, Italy 14; Grottes de Lascaux, Dordogne, France 75; private collection 4, 81, 96, 139a, 157; private collection/Archives Charmet 1204, 147, 186; private collection/© Aymon de Lestrange 206; private collection/Giraudon 22; private collection/© Look and Learn 142; private collection/Peter Newark Pictures 202; private collection/photo © Boltin Picture Library 219; private collection/photo © Bonhams, London, United Kingdom 187; private collection, photo © Christie's Images 107, 168; private collection/The Stapleton Collection 173, 210, 231; Deir el-Medina, Thebes, Egypt 237; Egyptian National Museum, Cairo, Egypt/AISA 78; Egyptian National Museum, Cairo, Egypt/photo © Boltin Picture Library 238; Galleria dell' Accademia, Venice, Italy 220; Graphische Sammlung Albertina, Vienna, Austria 90; Hamburger Kunsthalle, Hamburg, Germany 256; Herbert Ponting, Royal Geographical Society, London 28; Johnny van Haeften Gallery, London, United Kingdom 241; Leeds Museums and Galleries (City Art Gallery), United Kingdom 133; the Louvre, Paris 72; the Louvre, Paris, France/B. de Sollier & P. Muxel 127; the Louvre, Paris/Giraudon 38, 158; Mauritshuis 41; Musée de l'assistance publique, Hôpitaux de Paris, France 122; Musée des Beaux-Arts, Dijon, France/Giraudon 173; Musée des Beaux-Arts, Dijon, France/Peter Willi 229; Musée du Berry, Bourges, France/ Giraudon 144; Museo Correr, Venice, Italy 146; Museo della Specola, Florence, Italy 143; National Archaeological Museum, Athens, Greece/Giraudon 79; National Museum, Oslo, Norway 151; Prado, Madrid, Spain 11, 82; Service historique de la Marine, Vincennes France/Giraudon 73; Rijksmuseum Kroller-Muller, Otterlo, Netherlands 8; Saint-Pierre, Vatican, Italy 108; The Marsden Archive, United Kingdom 183; © Tokyo Fuji Art Museum, Tokyo, Japan 214, 215; Vatican Museums and Galleries 26; Walkert Art Gallery, National Museums Liverpool 67

Centers for Disease Control and Prevention: CDC/Dr. Fred Murphy 134b;CDC/Jean Roy 137; Janice Haney Carr 141

David M. Hillis, Derrick Zwickl, Robin Gutell (University of Texas): 45

Getty Images: 85, 129, 182, 185, 227; Aaron Graubart 13; AFP/Getty Images 30, 76, 130, 149, 156, 174, 203, 218, 223, 236; Andrew Errington 208; Anna Huerta 102; Anthony Bradshaw 207; Arctic-Images 24; Ashok Sinha 7; © Axel Lauerer 212; Bert Hardy 242; Celeste Romero Cano 152; Charles Thatcher 48; Christy Gavitt 46b; Comstock 155; David Becker 42, 119; David Fleetham/Visuals Unlimited, Inc. 164a; David Wrobel 164b; DEA Picture Library 239; Dr. David Phillips 50; Dr. Gopal Murti 64, 65, 226; Dr. James L. Castner 230; Dr. Kenneth Greer 118; Erik Dreyer 68; ERproductions Ltd 114; Frank Greenaway 232; Gordon Wiltsie 98; Henrik Sorensen 33; Ian Sanderson 180; Ihoko Saito/ Toshiyuki Tajima 46fn; James Balog 70; Jason Edwards 164d; Joe Raedle 37; John Sann 99; Kallista Images 148a; Kenneth Garrett 71; Luis Veiga 197; Mahaux Photography 193; Marco Di Lauro 192; Marcy Maloy 100; Mark Andersen 93; Mark Bolton 160; Mark Raycroft 229; Marta Bevacqua photos 52; Medioimages/Photodisc 46m; Michael Simpson 233; Moredun Animal Health Ltd./SPL 123; National Geographic/Getty Images 87; NHLI via Getty Images 196; OJO Images 106; Oxford Scientific 61; Oxford Scientific/Photolibrary 46d; PhotoAlto/Michele Constantini 177; Popperfoto/Getty Image 139b; Ralph Hutchings 61; Richard Ashworth 234; Sarah Faubus 51f; Science Photo Library 18; Simon Roberts 128; SSPL via Getty Images 205; Steve Allen 216; Steve Gschmeissner/SPL 115; Stocktrek Images 189; SuperStock 84; Time & Life Pictures/Getty Images 83, 94, 104, 136, 167, 170b, 178; Tim Flach 165; Todd Gipstein 40; Topical Press Agency/Stringer 49; Wood/CMSP 172

Groupe Librex: 25, 34, 97, 105, 221

Istockphoto: duncan1890 10

Jupiter Images: 93

Sarah Scott: 80, 89, 181b, 201, 253

Shutterstock: cover, 93, 188; A Cotton Photo 164c; Ajay Bhaskar 51g; Alexander Gitlits 131; Alleksander 51o; almondd 4650a; andesign101 110; Andrejs Pidjass 51b; ansar80 51h; Anyka 46k; Basov Mikhail 17; beaumem 161b; bikeriderlondon 138; Bliznetsov 202; Carolina K. Smith, MD, 56, 145; Carrie's Camera 46j; Carsten Reisinger 125; Cathy Keifer 46g; chuong 250, 251; Cigdem Cooper 86; delihayat 51j; Dmitry Savinov 228; Eastimages 77; EcoPrint 46ehiop, 166; ene 111; ggw1962 176; gorica 211; Graphic design 175; Gregory Johnston 51l; iDesign 23; iofoto 94; James Steidl 39; Jarno Gonzalez Zarraonandia 74; JinYoung Lee 103; Jose AS Reyes 46l; Kobby Dagan 51m; KUCO 1; lcepparo 51d; Izaokas Sapiro 93b; Levent Konuk 44; Lucian Coman 51c; margita 105; Martin Fowler 162; Melinda Fawver 163a; michaeljung 51a; Monkey Business Images 51k; Natale Matt 171; Natalia Sinjushina & Evgeniy Meyke 61; Olivier Le Queinec 217; ostill 51n; Pablo H Caridad 50; R. Gino Santa Maria 51i; Richards 224; Rob Marmion 51e; SFC 170a; Stanislav Bokach 53c; Stefan Schejok 58; Stephen Mcsweeny 190; Steve Smith Photography 46c; Steven Russell Smith Photos 163b; stocksnapp 161a; Stuart Monk 51p; Supri Suharjoto 92; Suzan Oschmann 244, 247, 249; Yuri Arcurs 95

Wikipedia: 204

Index